Practical Guide
to Designed
Experiments

MECHANICAL ENGINEERING
A Series of Textbooks and Reference Books

Founding Editor

L. L. Faulkner

*Columbus Division, Battelle Memorial Institute
and Department of Mechanical Engineering
The Ohio State University
Columbus, Ohio*

Additional Volumes in Preparation

Mechanical Engineering Software

Spring Design with an IBM PC, Al Dietrich

Mechanical Design Failure Analysis: With Failure Analysis System Software for the IBM PC, David G. Ullman

Practical Guide to Designed Experiments

A Unified Modular Approach

Paul D. Funkenbusch

University of Rochester
Rochester, New York, U.S.A.

CRC Press
Taylor & Francis Group
Boca Raton London New York

CRC Press is an imprint of the
Taylor & Francis Group, an **informa** business

CRC Press
Taylor & Francis Group
6000 Broken Sound Parkway NW, Suite 300
Boca Raton, FL 33487-2742

First issued in paperback 2021

© 2005 by Marcel Dekker
CRC Press is an imprint of Taylor & Francis Group, an Informa business

No claim to original U.S. Government works

ISBN-13: 978-0-8247-5388-7 (hbk)
ISBN-13: 978-1-03-218014-4 (pbk)
DOI: 10.4324/9780203997314

Library of Congress Cataloging-in-Publication Data
A catalog record for this book is available from the Library of Congress.

For Ming Tian, wife, partner, friend, and inspiration

Preface

Many graduating engineers and scientists receive little or no training in designed experiments. When coursework is provided it is often abstract and divorced from the practical considerations of interest in industrial application. One result has been the growth of a "short-course industry," which provides in-house training at the job site, often emphasizing the Taguchi method. These courses help practicing engineers and scientists get started with the techniques, but often lack the rigor and depth to develop true long-term understanding.

The objective of this book is to bridge this gap, by presenting the essential material in a fashion that permits rapid application to practical problems, but provides the structure and understanding necessary for long-term growth.

The book covers two-level and three-level full and fractional factorial designs. In addition, the "L12" and "L18" designs popularized by Taguchi are included. The role and selection of the system response for measurement and optimization are described. Both conventional and Taguchi ("S/N ratio") approaches are discussed and their similarities and differences described. Strategies for incorporating real world variation into the experimental design (e.g., use of "noise" factors) are presented and described. Data analysis by analysis of means, analysis of variance (ANOVA), and the use of normal probability plots will be covered.

The text presents the material using a modular or "building block" approach. In the first section a simple but complete design is presented and analyzed. Readers see how the entire structure fits together and learn the essential techniques and terminolgy needed to develop more complex designs and analyses. In the second section (Chapters 4–7), readers are taught more complex concepts and designs. The individual chapters are based on three essential building blocks: array design, response selection, and incorporation of "noise." The third section of the text (Chapter 8) deals with experimental analysis and follow-up in more detail.

The book is at a level suitable for those with a basic science or engineering background but little or no previous exposure to matrix experiments or other elements of planned experimentation. The primary audience is upper-level (junior and senior) undergraduates and first-year graduate students. The book can also be used as a self-study guide for scientists and engineers in industry.

Essential features of the book are:

Clear, factual instruction—explaining both "how" and "why"
Presentation of advanced techniques
Balanced discussion of alternative approaches
Examples and case studies from many disciplines
Homework/review problems

Paul D. Funkenbusch

Acknowledgments

I would like first to thank Mr. John Corrigan and the other good folks at Marcel Dekker for providing the opportunity and support necessary to make this book a reality. Professor Richard Benson was my department chair when I first became interested in the field of designed experimentation and provided me with a first chance to teach, explore, and develop an organizational structure for this material. Similarly, Profs. J.C.M. Li, Stephen Burns, and my other senior colleagues in the ME department at the University of Rochester provided the intellectual freedom and support necessary to further develop my approach.

Writing takes time. For this book that time came largely at the expense of my wife and children. They put up with my odd hours and (more than usually) eccentric behavior without quibble or complaint. I am deeply grateful.

Many people provided concepts and insight that helped in developing the approach presented here. In lieu of a long, and necessarily incomplete, listing, I can only express my sincere gratitude to the many colleagues, students, clients, and instructors who helped me to both understand the details and see the unifying structure. This book is my humble attempt to pass this understanding on—to present the beauty and power of these designs in a package that makes them available in a systematic and readily accessible form.

Contents

Building Blocks

Analysis of Results

1

How a Designed Experiment Works

Overview

This chapter provides an overview of how the components of an experimental design fit and work together. This is performed by walking through the design and analysis of a simple experiment. If you are not familiar with experimental design, this chapter will provide the necessary "road map" for placing the subsequent chapters (which describe different design and analysis components in detail) into proper context.

I. POLISHING FLUID DELIVERY SYSTEM

In recent years, the processing of precision optics has undergone a transformation from a craftsman-based process to one increasingly driven by computer numerical control (CNC) systems. Computer numerical control polishing relies on the generation of a consistent polishing "spot," so that the controller can predict the dwell time required to produce the desired removal and smoothing at each location on the optical surface. During preliminary testing of a new polishing machine prototype, concern was expressed over the consistency of the polishing fluid being supplied. In

1

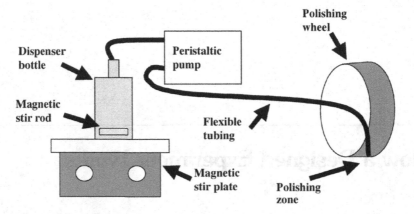

FIGURE 1   Schematic illustration of the polishing fluid delivery system.

particular, the concentration of abrasive supplied by the system was frequently much lower than the nominal value. An undergraduate student team [1] was recruited to develop a new fluid delivery system and to identify settings that could deliver concentrations close to the nominal. This example is primarily based on data collected during their work.

In building the prototype machine, an existing machine platform that had a coolant system, but no provision for including abrasives, was used. Therefore a fluid delivery system, capable of supplying an abrasive-containing polishing fluid mixture, had been externally added to the machine. Figure 1 illustrates the basic design of this external system. Polishing fluid of the desired composition is loaded into the bottle, along with a magnetic stirring bar. A magnetic stirring plate causes rotation of the bar and stirring of the fluid. Fluid is extracted from the bottle using a peristaltic pump and travels through a flexible plastic tube until it is deposited at the polishing zone.

II.   CONTROL FACTORS, LEVELS, AND CODING

The number of possible parameters for experimentation is often much larger than the number we can practically test in a single experimental design. Therefore the experimental team must do some preliminary winnowing, based on their assessment of likely importance. For the polishing fluid delivery system, the team identified four parameters for testing: bottle design, diameter of the supply tube, stir bar design, and flow rate through the tube. We refer to these as the control factors for this

Control Factor	Level	
	-1	+1
A. bottle geometry	convex	flat
B. tube diameter	3/32"	3/16"
C. stir bar geometry	pill	modified
D. flow rate	8 ml/min.	12 ml/min.

FIGURE 2 Control factors and levels for fluid delivery experiment.

experiment. In addition, the team was concerned about how the value of one control factor might affect the results obtained for another control factor, a phenomena referred to as an interaction. Two interactions were identified as being of particular concern—that between the design of the stir bar and the design of the bottle in which it moves, and that between the fluid flow rate and the diameter of the tube through which it flows.

To see the effect of a control factor on performance, it is necessary to change its value within the experimental design. For example, to see the effect of the bottle design, some experiments are run with one possible design and some with another. We then judge the importance of the design from the difference in the results between these two sets. The different values given to a control factor within an experimental design are referred to as its levels. For the fluid delivery system, two levels were selected for each of the control factors.

For convenience, it is useful to set up a simple shorthand system to refer to different factors and their levels. Control factors will be designated by capital letters (e.g., A, B) and levels by integers (e.g., -1, $+1$). Figure 2 shows this system applied to the fluid delivery system. Thus, for example, "Factor B, level -1" corresponds to using a tube diameter of 3/32 in.

III. ARRAY SELECTION AND SETUP

A crucial choice in experimental design is the selection and setup of the experimental array. For the fluid delivery experiment, the array selected is shown in Fig. 3. (This particular array is the 16 TC Factorial design from Appendix A, where TC stands for "treatment condition." To set up a specific experimental design using this array, each of the four factors is assigned to a specific column. You will learn how to create a design like this in Chap. 4). Although the array in Fig. 3 may look a little overwhelming at first, this is more a function of bulk than of complexity. The underlying concepts are quite simple, as can be seen by focusing on individual elements in the design.

TC	Columns														
	A	B	C	D	e	e	e	CxD	e	BxD	AxD	e	BxC	AxC	AxB
1	-1	-1	-1	-1	+1	-1	-1	+1	-1	+1	+1	-1	+1	+1	+1
2	-1	-1	-1	+1	-1	+1	+1	-1	+1	-1	-1	-1	+1	+1	+1
3	-1	-1	+1	-1	-1	+1	+1	-1	-1	+1	+1	+1	-1	-1	+1
4	-1	-1	+1	+1	+1	-1	-1	+1	+1	-1	-1	+1	-1	-1	+1
5	-1	+1	-1	-1	-1	+1	-1	+1	+1	-1	+1	+1	-1	+1	-1
6	-1	+1	-1	+1	+1	-1	+1	-1	-1	+1	-1	+1	-1	+1	-1
7	-1	+1	+1	-1	+1	-1	+1	-1	+1	-1	+1	-1	+1	-1	-1
8	-1	+1	+1	+1	-1	+1	-1	+1	-1	+1	-1	-1	+1	-1	-1
9	+1	-1	-1	-1	-1	-1	+1	+1	+1	+1	-1	+1	+1	-1	-1
10	+1	-1	-1	+1	+1	+1	-1	-1	-1	-1	+1	+1	+1	-1	-1
11	+1	-1	+1	-1	+1	+1	-1	-1	+1	+1	-1	-1	-1	+1	-1
12	+1	-1	+1	+1	-1	-1	+1	+1	-1	-1	+1	-1	-1	+1	-1
13	+1	+1	-1	-1	+1	+1	+1	+1	-1	-1	-1	-1	-1	-1	+1
14	+1	+1	-1	+1	-1	-1	-1	-1	+1	+1	+1	-1	-1	-1	+1
15	+1	+1	+1	-1	-1	-1	-1	-1	-1	-1	-1	+1	+1	+1	+1
16	+1	+1	+1	+1	+1	+1	+1	+1	+1	+1	+1	+1	+1	+1	+1

FIGURE 3   Array design and factor assignment for fluid delivery experiment.

Each numbered row in the array designates a particular treatment condition (TC). A treatment condition is a specific combination of control factor levels to be run (tested) during the experiment. Because there are 16 treatment conditions in this particular design, the experiment involves running 16 different combinations of control factor levels.

The numbered columns in the array have had one of three different types of designation added. Four of the columns are labeled with the control factor codes (i.e., A, B, C, D). These four columns provide the information actually necessary to conduct the experiment. In particular, they designate the settings of the control factors to be used for each treatment condition. For example, for TC 2, the A column has a value of −1. This means that factor A should be set to level −1 for this treatment condition. Referring back to Fig. 2, this means that a convex bottle should be used. Decoding the rest of the information for TC 2, it can be seen that the tube diameter should be 3/32 in. (B at level −1), the stir bar should have a pill geometry (C at level −1), and a flow rate of 12 ml/min (D at level + 1) should be used.

The remaining columns in Fig. 3 are not needed to perform the experiment, but will be used in analyzing the results. Six of the remaining columns (labeled A × B, A × C, A × D, B × C, B × D, and C × D) represent interaction terms between the various combinations of two factors. The designation on these columns means that we can use the coding in the column to help calculate the effect of the interaction when we analyze the experiment. The remaining columns in the design, labeled with the small letter "e," will be used to estimate the error in the experiment.

IV. NOISE FACTORS AND CONDITIONS

A common feature in many design and engineering problems is the presence of parameters that we believe are important, but that cannot be controlled in the field. It is important that such parameters be included in the experimental design so that recommendations based on the results will be valid under realistic conditions. For example, the environmental conditions (temperature, humidity) will influence the performance of an automobile starter. To ensure the starter works well when installed on customers' cars, it is, therefore, important that experimentation be conducted at different values (levels) of temperature and humidity. This type of parameter, which is varied in a controlled fashion during experimentation, but will be uncontrolled in the final application, is referred to as a noise factor. For experimental convenience, it is useful to have a designation to refer to specific combinations of noise factor levels. These combinations are designated as noise conditions (NC). Use of this terminology parallels the use of treatment condition (TC) to represent a specific combination of control factor levels. (Strategies for identifying noise factors and levels, and setting up noise conditions are discussed in Chap. 6). Extending our coding system to cover noise factors and levels, it is often helpful to use capital letters near the end of the alphabet (e.g., U, V, W) to designate noise factors and again designate levels in terms of -1 and $+1$. Noise conditions will be numbered sequentially (1, 2, 3, . . .).

For the fluid delivery system, concern was focused on the possibility that the fluid composition delivered might vary with the amount of fluid

	Noise conditions	
	1	2
Fluid pumped	10%	70%

FIGURE 4   Noise conditions for fluid delivery experiment.

TC	A	B	C	D	e	e	e	CxD	e	BxD	AxD	e	BxC	AxC	AxB	NC 1	NC 2	α
1	-1	-1	-1	-1	+1	-1	-1	+1	-1	+1	+1	-1	+1	+1	+1	$y_{1,1}$	$y_{1,2}$	α_1
2	-1	-1	-1	+1	-1	+1	+1	-1	+1	-1	-1	-1	+1	+1	+1	$y_{2,1}$	$y_{2,2}$	α_2
3	-1	-1	+1	-1	-1	+1	+1	-1	-1	+1	+1	+1	-1	-1	+1	$y_{3,1}$	$y_{3,2}$	α_3
4	-1	-1	+1	+1	+1	-1	-1	+1	+1	-1	-1	+1	-1	-1	+1	$y_{4,1}$	$y_{4,2}$	α_4
5	-1	+1	-1	-1	-1	-1	-1	+1	+1	+1	+1	+1	-1	+1	-1	$y_{5,1}$	$y_{5,2}$	α_5
6	-1	+1	-1	+1	+1	+1	+1	-1	-1	-1	-1	+1	-1	+1	-1	$y_{6,1}$	$y_{6,2}$	α_6
7	-1	+1	+1	-1	+1	+1	+1	-1	+1	+1	+1	-1	+1	-1	-1	$y_{7,1}$	$y_{7,2}$	α_7
8	-1	+1	+1	+1	-1	-1	-1	+1	-1	-1	-1	-1	+1	-1	-1	$y_{8,1}$	$y_{8,2}$	α_8
9	+1	-1	-1	-1	+1	-1	-1	+1	-1	+1	+1	+1	+1	-1	-1	$y_{9,1}$	$y_{9,2}$	α_9
10	+1	-1	-1	+1	-1	+1	+1	-1	+1	-1	-1	+1	+1	-1	-1	$y_{10,1}$	$y_{10,2}$	α_{10}
11	+1	-1	+1	-1	-1	+1	+1	-1	-1	+1	+1	-1	-1	+1	-1	$y_{11,1}$	$y_{11,2}$	α_{11}
12	+1	-1	+1	+1	+1	-1	-1	+1	+1	-1	-1	-1	-1	+1	-1	$y_{12,1}$	$y_{12,2}$	α_{12}
13	+1	+1	-1	-1	-1	-1	-1	+1	+1	+1	+1	-1	-1	-1	+1	$y_{13,1}$	$y_{13,2}$	α_{13}
14	+1	+1	-1	+1	+1	+1	+1	-1	-1	-1	-1	-1	-1	-1	+1	$y_{14,1}$	$y_{14,2}$	α_{14}
15	+1	+1	+1	-1	+1	+1	+1	-1	+1	+1	+1	+1	+1	+1	+1	$y_{15,1}$	$y_{15,2}$	α_{15}
16	+1	+1	+1	+1	-1	-1	-1	+1	-1	-1	-1	+1	+1	+1	+1	$y_{16,1}$	$y_{16,2}$	α_{16}

FIGURE 5   Addition of columns for noise conditions and characteristic response to experimental design for the fluid delivery experiment.

remaining in the bottle. To "capture" this possibility, it was decided to sample the fluid supplied near the beginning and near the end of each bottle used in testing. Specifically, the fluid was tested after 10% of the bottle had been pumped and again after 70%. Hence the amount of fluid already pumped was taken as a noise factor with two levels, 10% and 70%. With only one noise factor, there is no need to set up a full coding system of factors and levels for this example. Therefore information on noise conditions can be summarized in a single table as shown in Fig. 4.

Figure 5 shows the experimental design with the noise conditions added. Each noise condition is set up as a separate column on the right side of the matrix. Hence the noise conditions (columns) run perpendicular to the treatment conditions (rows), defining a 16 × 2 matrix of all TC/NC combinations. (The final column in this figure, α is discussed in the next section.)

Data is recorded for each TC/NC combination and the results recorded on the chart. This is shown in Fig. 5 using the coding $y_{i,j}$ to designate the result obtained with treatment condition i and noise condition j.

TC	Columns				NC		α_1
	A	B	C	D	1	2	
1	-1	-1	-1	-1			
2	-1	-1	-1	+1			
3	-1	-1	+1	-1			
4	-1	-1	+1	+1			
5	-1	+1	-1	-1			
6	-1	+1	-1	+1			
7	-1	+1	+1	-1			
8	-1	+1	+1	+1			
9	+1	-1	-1	-1			
10	+1	-1	-1	+1			
11	+1	-1	+1	-1			
12	+1	-1	+1	+1			
13	+1	+1	-1	-1			
14	+1	+1	-1	+1			
15	+1	+1	+1	-1			
16	+1	+1	+1	+1			

FIGURE 6 Table for use in setting up TC/NC combinations and recording data during experimentation.

Thus, for example, $y_{2,1}$ is the result recorded for TC 2 and NC 1; in other words, with a convex bottle (A at level $-$ 1), a tube diameter of 3/32 in. (B at level $-$ 1), a pill geometry for the stir bar (C at level -1), a flow rate of 12 ml/min (D at level $+$ 1), and 10% of the fluid pumped (noise condition 1).

Figure 6 shows a stripped down version of Fig. 5, with only the required control and noise settings for each measurement displayed. This type of table is useful for recording data during actual experimentation.

V. RESPONSES AND CHARACTERISTIC RESPONSES

To analyze the effects of the various control parameters, it is necessary to develop a number that characterizes the performance of the system for each set of control factors (i.e., for each treatment condition). This involves two important decisions: (1) what specific response (i.e., $y_{i,j}$) should be measured for each TC/NC combination and (2) how should the responses for the different NC be combined to give a single value that characterizes the performance for the TC. This single value is called the characteristic response for the treatment condition.

1. For the fluid delivery system, it was decided to focus on a single polishing solution, consisting of 9.8 wt.% 1-μm alumina powder in water. The response for each TC/NC combination (i.e., $y_{i,j}$) was measured as the wt.% of alumina in the solution actually delivered.
2. In preliminary testing, the main performance concern was the failure of the delivery system to deliver the nominal concentration, as a result of loss (settling) of alumina from the water. Oversupply (i.e., a concentration greater than the nominal) was much less of a problem. Therefore it was decided to initially focus on increasing the concentration supplied throughout the polishing process.

The average concentration was chosen as the characteristic response for analysis:

$$\alpha_i = \frac{y_{i,1} + y_{i,2}}{2}$$

where α_i is the characteristic response for treatment condition i. (You will learn more about how to identify and choose a response for measurement and a characteristic response for analysis in Chap. 5.)

Figure 7 shows the responses measured in the actual experiment along with the values of the characteristic response (α_i) calculated from

TC	A	B	C	D	e	e	e	CxD	e	BxD	AxD	e	BxC	AxC	AxB	NC 1	NC 2	α
1	-1	-1	-1	-1	+1	-1	-1	+1	-1	+1	+1	-1	+1	+1	+1	8.19	6.43	7.31
2	-1	-1	-1	+1	-1	+1	+1	-1	+1	-1	-1	-1	+1	+1	+1	2.56	1.94	2.25
3	-1	-1	+1	-1	-1	+1	+1	-1	-1	+1	+1	+1	-1	-1	+1	8.15	9.00	8.58
4	-1	-1	+1	+1	+1	-1	-1	+1	+1	-1	-1	+1	-1	-1	+1	8.30	8.87	8.58
5	-1	+1	-1	-1	-1	+1	-1	+1	+1	-1	+1	+1	-1	+1	-1	5.52	6.67	6.10
6	-1	+1	-1	+1	+1	-1	+1	-1	-1	+1	-1	+1	-1	+1	-1	6.61	4.43	5.52
7	-1	+1	+1	-1	+1	-1	+1	-1	+1	-1	+1	-1	+1	-1	-1	11.04	9.81	10.42
8	-1	+1	+1	+1	-1	+1	-1	+1	-1	+1	-1	-1	+1	-1	-1	8.98	9.36	9.17
9	+1	-1	-1	-1	-1	-1	+1	+1	+1	+1	-1	+1	+1	-1	-1	3.52	1.25	2.38
10	+1	-1	-1	+1	+1	+1	-1	-1	-1	-1	+1	+1	+1	-1	-1	1.94	1.10	1.52
11	+1	-1	+1	-1	+1	+1	-1	-1	+1	+1	-1	-1	-1	+1	-1	5.26	5.66	5.46
12	+1	-1	+1	+1	-1	-1	+1	+1	-1	-1	+1	-1	-1	+1	-1	6.95	7.08	7.02
13	+1	+1	-1	-1	+1	+1	+1	+1	-1	-1	-1	-1	-1	-1	+1	1.03	0.84	0.94
14	+1	+1	-1	+1	-1	-1	-1	-1	+1	+1	+1	-1	-1	-1	+1	5.56	1.32	3.44
15	+1	+1	+1	-1	-1	-1	-1	-1	-1	-1	-1	+1	+1	+1	+1	5.30	9.39	7.34
16	+1	+1	+1	+1	+1	+1	+1	+1	+1	+1	+1	+1	+1	+1	+1	5.44	7.31	6.38

FIGURE 7 Measured responses and values of the characteristic response for fluid delivery experiment.

them. To demonstrate how to compute α, we choose the responses produced for the second fluid delivery treatment condition:

$$y_{2,1} = 2.56 \text{ and } y_{2,2} = 1.94$$

These two responses are combined to give a single value of the characteristic response for the second treatment condition:

$$\alpha_2 = \frac{y_{2,1} + y_{2,2}}{2} = \frac{2.56 + 1.94}{2} = 2.25$$

This calculation is repeated for each value of i, producing 16 characteristic responses, one for each treatment condition. The calculated results are shown in Fig. 7.

VI. ANALYSIS OF MEANS

ANalysis Of Means (ANOM) is used to determine which level of each factor is the "best" and can also provide a relative ranking of their importance. As the name implies, ANOM involves comparing the mean (average) values produced by the different levels of each factor. Specifically, the characteristic responses for all treatment conditions, where the factor was at level -1, are averaged and compared to the average obtained with the factor at level $+1$.

Figure 8 shows this analysis applied to each of the columns in the fluid delivery experiment, with m_{+1} and m_{-1} defined as the averages for levels $+1$ and -1, respectively. Δ is the difference between m_{+1} and m_{-1} (i.e., $m_{+1} - m_{-1}$), which we will call the effect for the column. To illustrate these calculations, consider the column for Factor B (tube diameter). This column indicates a level of $+1$ for treatment conditions 5–8 and 13–16 and a value of -1 for treatment conditions 1–4 and 9–12. Thus

$$m_{+1} = \frac{\alpha_5 + \alpha_6 + \alpha_7 + \alpha_8 + \alpha_{13} + \alpha_{14} + \alpha_{15} + \alpha_{16}}{8} = 6.16$$

$$m_{-1} = \frac{\alpha_1 + \alpha_2 + \alpha_3 + \alpha_4 + \alpha_9 + \alpha_{10} + \alpha_{11} + \alpha_{12}}{8} = 5.39$$

and

$$\Delta = m_{+1} - m_{-1} = 6.16 - 5.39 = +0.77$$

These results are shown at the bottom of the column in Fig. 8 along with those for each of the other columns in the design. Figure 9 shows some of these same results graphically. Note first of all that the differences observed

TC	A	B	C	D	e	e	e	CxD	e	BxD	AxD	e	BxC	AxC	AxB	NC 1	NC 2	α
1	-1	-1	-1	-1	+1	-1	-1	+1	-1	+1	+1	-1	+1	+1	+1	8.19	6.43	7.31
2	-1	-1	-1	+1	-1	+1	+1	-1	+1	-1	-1	-1	+1	+1	+1	2.56	1.94	2.25
3	-1	-1	+1	-1	-1	+1	+1	-1	-1	+1	+1	+1	-1	-1	+1	8.15	9.00	8.58
4	-1	-1	+1	+1	+1	-1	-1	+1	+1	-1	-1	+1	-1	-1	+1	8.30	8.87	8.58
5	-1	+1	-1	-1	-1	+1	-1	+1	+1	-1	+1	+1	-1	+1	-1	5.52	6.67	6.10
6	-1	+1	-1	+1	+1	-1	+1	-1	-1	+1	-1	+1	-1	+1	-1	6.61	4.43	5.52
7	-1	+1	+1	-1	+1	-1	+1	-1	+1	-1	+1	-1	+1	-1	-1	11.04	9.81	10.42
8	-1	+1	+1	+1	-1	+1	-1	+1	-1	+1	-1	-1	+1	-1	-1	8.98	9.36	9.17
9	+1	-1	-1	-1	-1	-1	+1	+1	+1	+1	-1	+1	+1	-1	-1	3.52	1.25	2.38
10	+1	-1	-1	+1	+1	+1	-1	-1	-1	-1	+1	+1	+1	-1	-1	1.94	1.10	1.52
11	+1	-1	+1	-1	+1	+1	-1	-1	+1	+1	-1	-1	-1	+1	-1	5.26	5.66	5.46
12	+1	-1	+1	+1	-1	-1	+1	+1	-1	-1	+1	-1	-1	+1	-1	6.95	7.08	7.02
13	+1	+1	-1	-1	+1	+1	+1	+1	-1	-1	-1	-1	-1	-1	+1	1.03	0.84	0.94
14	+1	+1	-1	+1	-1	-1	-1	-1	+1	+1	+1	-1	-1	-1	+1	5.56	1.32	3.44
15	+1	+1	+1	-1	-1	-1	-1	-1	-1	-1	-1	+1	+1	+1	+1	5.30	9.39	7.34
16	+1	+1	+1	+1	+1	+1	+1	+1	+1	+1	+1	+1	+1	+1	+1	5.44	7.31	6.38
m_{+1}	4.31	6.16	7.87	5.48	5.77	5.05	5.44	5.98	5.63	6.03	6.34	5.80	5.85	5.92	5.60			
m_{-1}	7.24	5.39	3.68	6.07	5.78	6.50	6.12	5.57	5.92	5.52	5.21	5.75	5.70	5.63	5.95			
Δ	-2.93	0.77	4.19	-0.59	-0.01	-1.45	-0.68	0.41	-0.29	0.51	1.13	0.05	0.15	0.29	-0.35			

FIGURE 8 Calculated values of each column's effect for the fluid delivery experiment.

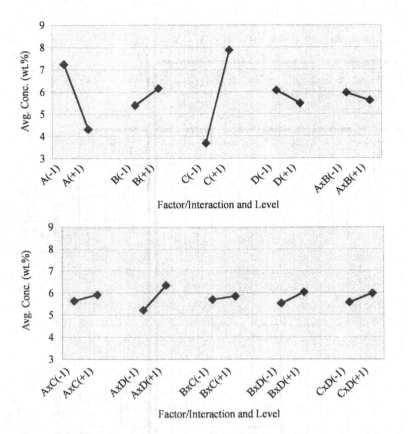

FIGURE 9 Graphical presentation of ANOM results (m_{-1} and m_{+1}) for the fluid delivery experiment.

for factors A and C are much larger than those for the other factors and interactions. So attention should be primarily focused on these two factors. Factor levels that produce larger values were desired because the goal was to increase the concentration of the fluid delivered by the system. Hence, from ANOM analysis, the "best" factor levels are determined to be −1 for factor A (convex bottle geometry) and +1 for factor C (modified stir bar). Settings for the other factors and interactions are less important.

VII. ANALYSIS OF VARIANCE

ANalysis Of VAriance (ANOVA) provides a second level of analysis. It is useful for judging the statistical significance of the factor and interaction effects observed. As the name implies, ANOVA involves study of the sources of variance in the data. In essence, a term proportional to the square of the differences revealed by ANOM analysis (i.e., $\Delta = m_{+1} - m_{-1}$) is divided by a similar term estimated for error to produce a ratio F. A large value of F indicates a variance much larger than that anticipated from error. Depending on the confidence level desired and some other details of the experimental design and analysis, a critical F can be defined. Factors or interactions that produce F values larger than the critical F value are judged statistically significant.

For the fluid delivery experiment, it was decided to use a 90% confidence level to judge significance, which gave a critical F value of 4.06. Figure 10 summarizes the results of ANOVA for the experiment. This table is written in standard format that includes information not needed at present. (Details of this and other aspects of both ANOM and ANOVA analysis are presented in Chap. 3.) However, focusing on the last column, we can obtain the central findings of the ANOVA analysis.

All of the interactions and two of the factors (B and D) have F values much lower than the critical value of 4.06. This does not necessarily mean they do not influence the concentration delivered by the supply system. However, if they did have an effect, it was too small to be resolved against the background of experimental error. Two of the factors (A and C) have F values much larger than the critical F value, and are judged significant.

Source	SS	DOF	MS	F
A. (Bottle geometry)	34.37	1	34.37	**16.1**
B. (Tube diameter)	2.41	1	2.41	1.1
C. (Stir bar)	70.10	1	70.10	**32.9**
D. (Flow rate)	1.35	1	1.35	0.6
AxB	0.48	1	0.48	0.2
AxC	0.35	1	0.35	0.2
AxD	5.21	1	5.21	2.4
BxC	0.08	1	0.08	0.04
BxD	1.04	1	1.04	0.5
CxD	0.70	1	0.70	0.3
error estimate	10.64	5	2.13	-
Total	126.828	15	-	-

FIGURE 10 ANOVA table for the fluid delivery experiment.

	Concentration (10% of fluid pumped)	Concentration (70% of fluid pumped)	α
optimized (off-line)	9.64	9.02	9.33
	9.41	11.16	10.28
optimized (on system)	9.26	9.13	9.20
	10.75	9.50	10.12

FIGURE 11   Results of subsequent testing of the fluid delivery system using the recommended settings.

These results are very similar to those obtained from the ANOM analysis. In particular, two of the factors are singled out from the others as "important." What is added by ANOVA is an assessment of whether these effects are "real" (statistically significant) or not. In particular, for the fluid delivery system, the ANOVA analysis provides confidence that the selection of the proper levels of factors A and C really will aid in increasing the concentration of fluid supplied.

Tests of the fluid delivery system were run with the recommended settings both off-line and installed on the polishing system. Results are summarized in Fig. 11. There is clearly still room for improvement. However, the system was able to consistently deliver a polishing fluid with a concentration near the 9.8% nominal. (Compare these results with the range of values seen in Fig. 8.)

VIII.   CONCLUSION

This chapter provided, by way of a working example, an introduction to the key elements of practical designed experiments and a perspective on how they fit together to make a complete experiment. These key elements include design of the experimental array, selection of a suitable response and characteristic response, incorporation of real-world variability, and statistical analysis of the results. The following chapters provide more detailed information on these key elements, both detailed instructions and "libraries" of alternative approaches and strategies.

Chapter 2 provides basic background on contemporary product design and quality concepts, important for properly understanding the approaches and strategies described in the following chapters.

Chapter 3 explains how to analyze designed experiments and interpret their results, including a more detailed explanation of ANOM and ANOVA.

Chapter 4 deals with experimental array design for factors with two levels. It begins with a description of the arrays and discussion of their fundamental trade-offs and limitations. It then explains how to set up arrays to match specific applications, including a library of basic experimental design strategies.

Chapter 5 describes how to select and refine a response and characteristic response to meet specific experimental and project goals.

Chapter 6 deals with incorporation of real-world (field and manufacturing) variability into the experimental design. This includes discussion of different types of "noise factors" and a library of strategies for incorporating them into an experiment.

Chapter 7 continues discussion of experimental array design, extending it to arrays with three-level factors, as well as explaining how arrays can be modified to include factors with "nonstandard" numbers of levels.

Chapter 8 extends discussion of how to analyze designed experiments and interpret their results, by presenting several additional techniques. These include pooling of smallest effects, replication, and use of normal probability plots.

2

Fundamental Concepts

Overview

In this chapter, you will gain some basic context on the role and uses of designed experimentation for solving practical problems. Much of the layout and analysis that you will see in subsequent chapters is motivated by "philosophical" considerations such as project organization, design goals, and the definition of "quality." Gaining an understanding of these is important in understanding why particular definitions and structures are used. It will also guide you in making rational choices as you develop experimental designs to fit your own projects' specific goals.

I. FOOD FOR THOUGHT

The sign shown in Fig. 1 appeared on my department's copier several years ago. (The sign is reproduced verbatim except that, for obvious reasons, the copier maker's name has been deleted.) Please read it and then, before reading the subsequent discussion, consider for a few seconds how you might respond to each of the following questions.

16

Important Notice

When using copier after 5 p.m.
Please make sure that you shut-
off the copier and close the copier
room door before you leave the
building.

The copier has been over heating
due to the fact that the copier is
being left on.

↳ or Poor Design!

No - copier not meant for
Such a small room - give
the
engineer a have
credit!

FIGURE 1   Customer feedback. Who is to blame?

A.   Is this a Case of Poor Design? Improper Use? Who is at Fault?

Discussion of who's at fault can grow quite contentious. At various times, I
have heard blame placed on the design team, the factory, the sales
representative, and the customer. A relatively common theme is to look
for written requirements for the room size or airflow in the owner's
manual. If the room is "big enough" according to these, then the copier
must be at fault. If, however, it is even a single cubic foot under the
requirement, then it is the customer's fault. Fundamentally, of course, such
arguments miss the most important point. From a customer's perspective,
a copier that does not work well is unacceptable regardless of whether it
reflects a problem with design, or manufacturing, or the environmental
conditions (room). This is unlikely to change based on the fine print in the
manual. Conversely, a customer who placed an identical copier in a less
stressful environment might be satisfied with its performance even if the
copier is sensitive to overheating. The performance obtained from a
product, and therefore its perceived quality, depends both on the product
itself and on the conditions under which it is used.

B.   Is this a Poor Quality Machine?

This question really asks you to make a false choice. Bad quality or not? In
the first place, we have not defined quality in a meaningful way. Moreover,
the situation we are considering is clearly more complex than can be
accounted for by a simple yes or no answer. This particular machine works

well most of the time (during the day) but needs some extra care at night. Clearly, it could be better, but it also might be worse (e.g., if it overheated during the day). Different customers (with different environments) might also have very different experiences. In discussing quality, therefore, we need to be specific in our definition and to consider the influence of variability on performance.

C. How Would You Go About Fixing the Copier? How Would You Go About Designing an Improved Copier?

Responses as to how to fix a problem are generally more positive and creative than those involving assignment of blame. Possible remedies include fans, heat pumps, insulation, providing a lower energy "sleep" mode, and improving the energy efficiency of the copy process (which uses heat to permanently fuse the image onto the paper). In general, two broad categories can be distinguished: those that treat the symptom (accumulation of heat) and those that treat the source (the amount of heat produced). In dealing with an existing problem, treating the symptom, sometimes, is an effective approach. For example, it might be easier to retrofit a fan into the above copier than to try to install a completely new fuser system. In product design and manufacturing, however, an emphasis on the fundamentals (sources) is often most valuable. Care should be taken to avoid letting the design process be dominated by the practice of identifying and "fixing" specific problems, at the expense of making fundamental improvements in function and efficiency.

II. OVER-THE-WALL ENGINEERING

Machinists, especially those who work in academia, get to see many interesting blueprints: solid blocks with spherical cavities somehow machined inside, threads in the form of a mobius loop, and so on. These are extreme examples of a generic problem, a tendency to overcompartmentalize design and manufacturing processes. Figure 2a illustrates this problem. The design division passes its blueprint "over-the-wall" separating it from manufacturing, relying on them to somehow output the product as designed. Figure 2b shows a related problem, in which a product designed in a laboratory environment is passed "over-the-wall" to the customer, where it must somehow be made to function in a very different environment.

Manufacturability

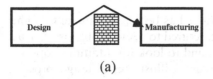

(a)

In service performance

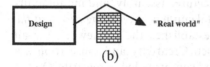

(b)

FIGURE 2   "Over-the-wall" design.

A wide range of skills is necessary to bring a complex product from the design board to the market. So it is reasonable to organize the process in terms of divisions/fields such as design, manufacturing, and customer service, with expertise in specific areas. The problem is, therefore, not with the compartmentalization itself. It is with how well information about the manufacturing and operating environments can be passed "upstream" (from "the field" to testing to manufacturing and then to design) and with how that information can best be incorporated into design decisions.

III. CONTRIBUTIONS OF TAGUCHI

Genichi Taguchi (1924–) has played a major role in the field of quality engineering and improvement, including both the development of experimental design and in its application to solving practical problems. His work has stimulated a great deal of thought and analysis and there have been some sharp controversies. However, even those critical of some specifics in his approach generally recognize the fundamental importance of the underlying concepts. We will be learning from and interpreting Taguchi's work in many places in this chapter and throughout this book. In the following sections of this chapter, some of the more important of the concepts introduced by Taguchi are described. In subsequent chapters, we will use these ideas to help guide the development of experimental designs.

IV. STAGES OF DESIGN

One of Taguchi's important contributions has been to stimulate thought about how the design process can be structured to avoid problems, such as those inherent in over-the-wall design, and to look for additional opportunities for product improvement. Figure 3 illustrates a design process based on some of his ideas. It consists of three distinct stages: concept, parameter, and tolerance design.

In *concept design*, the basic architecture, assembly, and manufacturing process are defined. The fundamental emphasis is *making sure that the underlying physics works*. This stage establishes the framework for all subsequent development of the product. Creativity and innovation are often important parts of this process and can provide big rewards. On the other hand, mistakes may be difficult, impossible, or costly to correct at later stages. For example, we might choose an internal combustion, electric, or hybrid power plant for a vehicle in the concept design stage. Subsequent design can be used to modify the particular choice made, but the initial concept selection will largely determine the design space (e.g., range of weight, fuel efficiency, pollution, and cost) in which they can work.

In *parameter design*, the nominal values of key specifications are selected so that the product works consistently well in spite of the variability inherent in the manufacturing and/or operating environments. The fundamental emphasis is on *making the physics work under real-world conditions*. For example, by matching the thermal expansion coefficients of parts in intimate contact, a design less sensitive to operating temperature can be produced. These types of choices often have only a minor impact (positive or negative) on production costs, but can lead to major improvements in final performance.

In *tolerance design*, practical trade-offs are made between performance and manufacturing cost. The fundamental emphasis is on *making the economics work*. As the name implies, tolerance design often involves determination of the manufacturing tolerances about the specified nominal values. The tighter the tolerances, the closer the actual product will be to

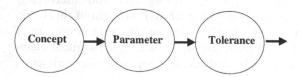

FIGURE 3   Three-stage design process.

the ideal specified in the blueprints, but the higher the cost. So components must be prioritized so that tolerances are tightened where they will produce the most impact for the least cost.

To illustrate these stages, consider the design of a simple mechanical device such as a catapult. At the end of the concept design stage, the basic physics will have been worked out and there will be a drawing of the device, but many specifications will still be "open." For example, a range of possible launch angles may have been identified, but the specific angle may still be open (launch angle $= \Theta$). At the end of the parameter design stage, the drawing would include nominal dimensions and other specifications (launch angle $= 30°$). At the end of the tolerance design stage, tolerances are added to the drawing producing a blueprint suitable for assembly/manufacturing (launch angle $= 30 \pm 5°$).

In a large organization, these different stages could be different boxes on an organizational chart, with different scientists and engineers assigned to each. In a small organization, the same few people may be involved in each of the stages. The important point is not the formal structuring, but the push to look at a design problem in several distinctly different ways. In particular, parameter design is a process that can deliver large benefits, but is often neglected. Without parameter design, specifications may be set at the end of the concept design stage and passed through for tolerancing, without the chance to experimentally test alternatives or to evaluate the effects of variability in the manufacturing and operating environments.

Designed experiments are of particular importance in successful implementation of parameter design, and many of the examples you will see in this textbook may be thought of as parameter design problems. Designed experiments are also important in tolerance design, specifically to experimentally determine which manufacturing tolerances have the largest effects on the consistency of product performance. Concept design, which generally involves choices among fundamentally different design paths, is less amenable to this type of experimental analysis.

V. DEFINING QUALITY (TRADITIONAL APPROACH)

In the everyday world, "quality" is often used as a synonym for "good" or "nice." We are sold "high-quality" electronics by "quality" employees who do only "good quality" work. This may be good public relations, but it does not tell us very much about what is really happening.

A more meaningful approach to quality begins by considering the cost effects of having a component (or product) that does not exactly match its ideal specifications/requirements. As a simple example, consider a metal pin that is designed to fit snuggly into a matching hole during product assembly. There is an ideal diameter for this pin, for which the assembly works perfectly as designed. If, however, the pin diameter deviates from this value, it will not work as well. If it is too large, it will be difficult (or impossible) to insert and may cause unwanted friction and wear. If it is too small, it will be wobbly or even fall out. As a measure of the quality, we define the loss, L, as the dollar cost of having an actual value, y, which deviates from the ideal value, m. This can be expressed in equation form as follows:

$$L(y) = f(y - m) \tag{1}$$

where f stands for "is a function of." Note that when $y = m$, the quality is perfect so there should be no loss and $L = 0$. The important part of the definition is, however, what we assume happens when y moves away from m.

Historically, industry has tended to look at quality in terms of the tolerances needed to assemble a product from mass-produced parts. Each component is given an ideal specification (m) and a plus/minus tolerance (Δ). For example, the diameter of the pin might be specified as 1 ± 0.01 cm. An individual part within these limits (i.e., between 0.99 and 1.01 cm in diameter) gives an acceptable result and can be used. A part outside of these limits (< 0.99 or > 1.01 cm) does not produce an acceptable result and should be rejected. Expressing this in equation form gives:

$$L(y) = 0 \qquad m - \Delta < y < m + \Delta$$
$$L(y) = k_s \qquad y < m - \Delta, y > m + \Delta \tag{2}$$

where k_s is a constant reflecting the cost of remanufacturing (scrapping and replacing) the part.

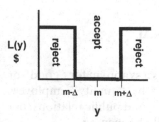

FIGURE 4 "Traditional" pass/fail system for assessing quality.

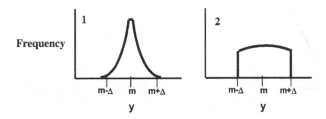

FIGURE 5   Part distribution from two possible suppliers.

This pass/fail (accept/reject) system for judging quality produces a step function for L and is illustrated in Fig. 4. It ties quality directly to the traditional engineering use of tolerances and provides a relatively simple way of estimating the cost associated with having some distribution of imperfect parts. However, defining quality this way also has its drawbacks. Most fundamentally, this system provides a poor representation of actual performance as a measure of quality. A part which only marginally within tolerances is rated equally with one that has exactly the desired dimensions, although their performance may differ markedly. Moreover, this same marginally in tolerance part is rated completely differently from one that is only marginally out of tolerance, although their performance is probably indistinguishable.

Beyond the seeming "unfairness" of this system, the poor tie to performance can have other negative consequences. Consider two-part suppliers, both of whom deliver equally high percentages "within tolerance," as illustrated in Fig. 5. The first supplier accomplishes this with a tightly controlled process, whose distribution is sharply spiked at the optimum value, m. The second supplier has a poorly controlled process with a wide distribution, but gets the necessary percentage within tolerance by screening out parts beyond the allowable limits. In terms of the pass/fail system, they will be ranked equally. However, clearly, the first supplier is delivering a superior quality product.

VI.   QUADRATIC LOSS FUNCTION

Recognizing some of the problems with this traditional approach to defining quality, Taguchi proposed the *quadratic loss function* as an alternative. If a part has a value that is different from the ideal (i.e., $y \neq m$), its performance will be inferior even if it is "within tolerance." The cost of this reduced performance may be considered a measure of the quality loss.

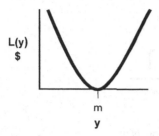

FIGURE 6   The quadratic loss function for assessing quality.

The quadratic loss function assumes that this loss scales with the square of the distance from the ideal. Thus

$$L(y) = k_q(y - m)^2 \tag{3}$$

where k_q is a constant. This relationship is shown in Fig. 6.

The selection of a quadratic function can be rationalized in different ways, but obviously is somewhat arbitrary. One nice feature of Eq. (3) is that it produces a very simple form when applied to a distribution of parts. Averaging the loss from Eq. (3) over a distribution with an average of μ, and a standard deviation of σ, gives:

$$Q = k_q\left[(\mu - m)^2 + \sigma^2\right] \tag{4}$$

where Q is the average quality loss (cost) per part. This formula emphasizes the fact that both average and variability are important in determining the quality of a product. Applying this formula to the case shown in Fig. 5, for example, the superior performance expected from the first supplier's product would be reflected in a lower value of Q, as a result of a smaller value for the σ^2 term.

Finally, note that, while we have so far discussed quality in terms of variability in the product, the same concepts are applicable to variability in product performance caused by variations in operating environment. In this case, σ would reflect the spread in performance caused by the environment.

VII. THE P-DIAGRAM

The P (product or process) diagram is a final conceptual tool, useful for understanding how performance and environment are related. Figure 7

Ideal

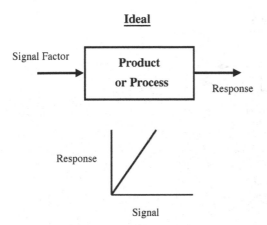

FIGURE 7   Ideal response of product or process—response follows signal factor with desired slope.

shows the first step in constructing a P-diagram. Under ideal conditions, we input a signal factor(s) to the product and output a corresponding response. A *signal factor* is a parameter set by the end user to control the product performance. As shown in the graph, under ideal conditions, there is a one-to-one correspondence between the selected value of the signal (input) and the response (output). Signal factors are most easily understood in terms of literal signals, such as the setting on a knob or the voltage on an electronic controller. However, it is also sometimes helpful to think in terms of signal factors in terms of physical inputs to the system, such as power supplied or raw material flow. In this case, the ideal signal–response relationship represents the successful conversion of this input into final form/product, e.g., conversion of input power into motion (kinetic energy).

In Fig. 8, the effects of variations in the product or the operating environment are introduced in the form of noise factors. A *noise factor* is defined as a parameter that is left uncontrolled or cannot be controlled in actual operation. Therefore any influence of noise factors on the response results in an undesirable variation when the product is actually used. As shown in Fig. 8, random scatter in the response is introduced and the slope of the signal/response curve may be changed.

Finally, Fig. 9 shows the completed P-diagram, in which control factors have been selected to optimize the design. A *control factor* is a parameter that is tested during the design process in order to identify its best value. This value is then fixed in the final design. As shown in the

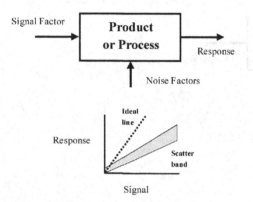

FIGURE 8 Uncontrolled (noise) factors can distort the response, changing the slope from the ideal value and causing scatter.

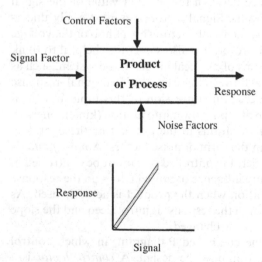

FIGURE 9 Completed P-diagram, illustrating the role of control factors in restoring the desired correspondence between signal and response.

graph, proper selection of control factors is used to reduce the variability in the response and adjust the slope of the signal/response curve onto the desired value.

To illustrate use of the P-diagram, consider its application to the design of an industrial furnace. The desired response for the furnace is the internal temperature, which is used in heat-treating metal parts. The signal factor is simply the setting on the temperature controller (or, possibly, the power input to the heating elements). Ideally, the temperature everywhere inside the furnace would correspond identically to the value set by the temperature controller. However, we know that the actual temperature in the furnace may be influenced by a wide variety of parameters that are difficult or impossible to control in service. These constitute the noise factors for the furnace. Examples include the external temperature and humidity, the power supply voltage, the number and size of parts (thermal load) placed in the furnace, the location of individual parts within the furnace, and the aging of the heating elements. The design engineer needs to be aware of these parameters and their potential influence when he or she selects specific values for the parameters that can be specified. These parameters that the design engineer can specify represent the control factors. Examples for the furnace include the type, number, and layout of the heating elements, the number and location of the temperature sensors, the controller (e.g., on/off vs. proportional, response time, etc.), and the shape and size of the chamber.

VIII. CONCLUSION

This chapter introduced some important current concepts in the fields of manufacturing, quality, and design, including the need to avoid "over-the-wall" engineering, the concept of parameter design, the importance of variability in assessing "quality", and the use of the P-diagram for visualizing the roles of noise and control factors in determining performance. These concepts have played an important role in the development of the overall framework for experimental design in industry. Their influence on many specific aspects of the experimental design process will be seen in the following chapters.

Homework/Discussion Problems

1) Describe a recent product or process "failure" that you have either witnessed or been involved in (e.g., your car not starting). Who do

you feel was to blame for the failure? Do you think it was an example of "over-the-wall" design? Explain why or why not.

2) Describe an example of something you feel represents "over-the-wall" design, preferably something from your own experience. Be specific about what seemed to have been missing in the design process.

3) Describe how use of a three-stage design process as described in this chapter (Fig. 3) can help avoid problems with "over-the-wall" design.

4) Assume that you have been assigned to the design team for a new coffeemaker. Describe the major decisions that you would need to make during each of the three design stages shown in Fig. 3. Be specific.

5) Many manufacturers, particularly small ones, still prefer to use strictly pass/fail design specifications to determine the acceptability of parts. For example, they want the customer to specify his or her requirements strictly in terms of a part size and fixed tolerance—such as 2 ± 0.02 cm.

    a. Explain, from the standpoint of such a manufacturer, why he or she may prefer such a system.
    b. What advantages does this system also offer to you, the customer?
    c. Describe some of the incentives for the customer to change this system (i.e., what are some of the problems and limitations).

6) Refer to Fig. 5. Manufacturing the distribution shown for supplier 2 (right side) may actually be more costly than that the tighter distribution shown for supplier 1. For example, consider the need to "screen" parts to obtain an acceptable distribution. Describe at least two ways that such screening could raise costs.

7) The quadratic loss function [Eq. (4)] emphasizes the roles of both the average and variance in determining quality. Advocates sometimes apply this equation directly to product data as a way of making quantitative comparisons. Skeptics generally accept the concept behind the quadratic loss function, but do not believe it should be interpreted literally.

A manufacturing process for powder metallurgy parts is designed around an ideal linear shrinkage of 8.9%. (Both higher and lower shrinkages are possible but undesirable.) Sample powders supplied by two different vendors give different results when used for this process.

Vendor 1: 8.9% average shrinkage with a standard deviation of 1.5%.

Vendor 2: 9.1% average shrinkage with a standard deviation of 0.25%.

    a. Apply the quadratic loss function to determine which vendor's powder produces better results.
    b. How would a skeptic view this assessment and why?
    c. Propose an alternative way to make this comparison. Contrast your proposal with the quadratic loss function (advantages and disadvantages).

    8) Draw the "P-diagram" and use it to explain the relationship between control factors and noise factors.

    9) Select a product from your home or work (e.g., a refrigerator). Draw a P-diagram for it and identify a signal factor, response, and at least five control and noise factors.

    10) Clearly and concisely distinguish between the following types of "factors." (That is, how would you determine the category to which a factor belongs?)

    a. Noise factor vs. control factor
    b. Signal factor vs. the response
    c. Signal factor vs. control factor

    11) In analyzing the response of a car, the extent to which the accelerator is depressed is modeled as a "factor."

    a. Give an example of a response for which the depression of the accelerator would be a *signal factor*. Explain.
    b. Give an example of a response for which depression of the accelerator is best considered a *noise factor*. Explain.

3

Statistical Concepts

Overview

This chapter is divided into two sections. In the first, some basic statistical concepts are reviewed and illustrated. In the second, the procedure for using analysis of means (ANOM) and analysis of variance (ANOVA), which you explored in application in Chap. 1 is developed in more detail. Upon successfully completing this chapter you will be prepared to perform ANOM and ANOVA and interpret the results from these analyses.

I. REVIEW OF BASIC CONCEPTS

A. Purpose of Review

Most engineers and scientists have had some exposure to basic statistics, such as calculating a standard deviation or computing a confidence limit with a normal distribution. In contrast, relatively few have performed an analysis of variance, which may appear quite complex in contrast. In fact, however, ANOVA is based on many of the same fundamental principles as the more familiar techniques. This section provides a quick overview of some basic statistics, including many points useful in understanding analogous operations in ANOM and ANOVA.

B. Distribution

Any real measurement contains some degree of variability, either as a consequence of real differences in the measurand (the quantity being measured) or uncertainty in the measurement system. For example, a manufacturing process may be designed to produce sections of plastic pipe 1 m in length, but the measured length of any individual section will undoubtedly deviate somewhat from this specification.

The variability in measured values is most conveniently captured using a frequency distribution such as those illustrated in Fig. 1. The horizontal axis shows the measured value, while the vertical axis defines the relative probability that a given measurement will lie within a certain range. For Fig. 1a, the possible measured values are sorted into discrete "bins," so that a discrete distribution is shown. The probability that a randomly chosen measurement is within the range defined by a particular bin may be read off the vertical axis directly. Alternatively, one can say that if a large number of measurements are taken, the vertical axis gives the fraction expected to be within each bin (i.e., the frequency of occurrence). Note that the sum of frequencies for all bins must add up to 1.

Figure 1b shows a continuous distribution. In this case, the "width" of each bin is infinitesimally small, so the probability of having exactly some value is also vanishingly small. Instead, it is necessary to think in terms of the probability of having a measured value within a specified range (in effect to define a "bin" so that the probability may be determined). This is given by the area under the curve between the specified values and can be obtained by integrating the (equation for the) curve between the limits specified for the range. Note that in this case the total area under the curve must again be equal to one.

The complete frequency distribution curve fully "captures" information about the distribution of the measured value. In practice, however, it is common to summarize this information in simpler forms. This may be done because full information is lacking or simply for convenience. A retail purchaser of the plastic pipe, e.g., would probably be more confused than enlightened by a frequency distribution curve. Average length is probably enough for the purchaser. On the other hand, the manufacturer probably needs to track variability. In product manufacturing, the complete frequency distribution, including the shape, is of interest.

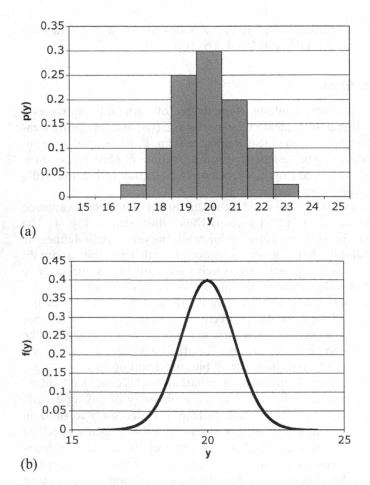

FIGURE 1 (a) Discrete and (b) continuous distributions.

C. Position

The average, or mean, is the sum of the individual measurements divided by the total number of measurements. It can be found from the following formula:

$$\bar{x} = \frac{\sum\limits_{i=1}^{n} x_i}{n} \tag{1a}$$

where $\bar{x}$ is the average for the sample, x_i are the individual measured data points, and n is the total number of data points in the sample. This is the most common way to quantify curve position, and generally the most mathematically useful.

If data from a complete population are used, the corresponding equation is:

$$\mu = \frac{\sum_{i=1}^{n} x_i}{N} \tag{1b}$$

where μ is the true average for the population, x_i are the measured individual data points, and N is the total number of data points in the population. $\bar{x}$ is an estimate of the true population value, μ.

Alternatives to using the average do exist. The median is the value for which the number of measurements below the median is equal to the number of measurements above it. Since the median is based only on balancing numbers above and below it is not as distorted by extreme values as the average, a property that is sometimes very useful. For example, a single billionaire moving into a city of 1 million people would raise the city's average net worth per person by more than $1000, but have essentially no affect on the median. Finally, the position is also sometimes quantified in terms of the mode(s), defined as the value(s) at which the frequency curve reaches a maximum. This is easy to do graphically and can be useful in identifying subpopulations.

D. Dispersion

Dispersion is a measure of the scatter about the mean. The simplest measure of dispersion is the range, defined as the difference between the maximum and minimum values measured. Range is sometimes used in statistical process control applications and to provide error bars on experimental data. However, it is difficult to use quantitatively and subject to distortion by outliers.

Standard deviation is one of the most useful measures of dispersion. Standard deviation is often estimated from a sample of the total population by using the following formula:

$$s = \sqrt{\frac{\sum_{i=1}^{n}(x_i - \bar{x})^2}{n - 1}} \tag{2a}$$

where s is an estimate of the true (population) standard deviation based on a finite sample of data, x_i are the individual data points, $\bar{x}$ is the average of the sample, and n is the number of data points in the sample. If, instead of a sample, all members of the population are used, the true (population) standard deviation can be obtained as:

$$\sigma = \sqrt{\frac{\sum_{i=1}^{N}(x_i - \mu)^2}{N}} \tag{2b}$$

where σ is the true (population) standard deviation, μ is the population average, and N is the number of points in the population. From an engineering standpoint, standard deviation is a convenient measure of dispersion because it has the same units as the value being measured.

Variance is given by the square of the standard deviation:

$$s^2 = \frac{\sum_{i=1}^{n}(x_i - \bar{x})^2}{n-1} \tag{3a}$$

$$\sigma^2 = \frac{\sum_{i=1}^{N}(x_i - \mu)^2}{N} \tag{3b}$$

One key advantage of variance is that uncorrelated (independent) variances can be added, whereas standard deviations cannot. To illustrate, consider a large distribution of plastic pipes 1 m long with a standard deviation of 1 cm. If these pipes are randomly selected and placed together (end to end) in sets of two, what is the average and standard deviation of the resulting distribution?

The average is clearly 2 m but the standard deviation is not 2 cm. Instead, the standard deviation must be calculated by first adding the variances and then converting to the standard deviation. In this case, the variance of the new distribution is $(1 \text{ cm})^2 + (1 \text{ cm})^2 = 2 \text{ cm}^2$. The standard deviation is the root of the variance, $\sqrt{2} \text{ cm} \approx 1.4 \text{ cm}$.

Note that this type of calculation only works if the variances are uncorrelated. In this example, this was assured by stating that the total number of pipes is large and the selection of two pipes to be placed together is random. Hence, the distribution of lengths from which the second pipe is selected is not influenced by the selection of the first pipe. If we tried to place large and short pipes together to "compensate" for each other, this calculation would be invalid. Being able to add (uncorrelated) variances is very important in tolerancing. It also plays a key role in statistical analysis.

One additional measure of the dispersion that is often encountered is the coefficient of variance, C_v. This is simply the standard deviation divided by the average. It is natural to expect tighter distributions, in absolute terms, on smaller parts than on larger parts. For example, a 1-mm standard deviation is quite large for a part with an average length of 1 cm, but seems small for a 1-m-long part. The coefficient of variance "normalizes" the dispersion against the average to obtain a relative measure:

$$C_v = \frac{\sigma}{\mu} \approx \frac{s}{\bar{x}} \tag{4}$$

E. Degrees of Freedom

The number of impendent quantities that can be calculated from experimental data cannot exceed the number of data points. Degrees of freedom (dof) are a useful way of accounting for the amount of data available, and used, in analyzing an experiment.

For example, in Eq. (3a), the square of the difference between a data point and the average may be thought of as an estimate of the true (population) variance. In the numerator of Eq. (3a) the terms for each of the data points are summed. In the denominator, this sum is divided by the number of independent estimates involved. Although there are n data points, these same points were used to calculate the estimate of the average value ($\bar{x}$), which also appears in Eq. (3a). Therefore, there are only $n - 1$ independent estimates of the variance.

F. The Normal Distribution

Figure 2 shows the familiar "bell-shaped" curve associated with a normal distribution. A normal distribution is often observed experimentally and also often assumed in statistical analyses. Why should this be so? If a measured value reflects the sum of a large number of small random events it turns out that it will tend to be distributed normally, regardless of the distribution associated with each of the individual events.

A simple example is useful to illustrate this effect. If we roll a standard (fair) six-sided die, the frequency distribution of the outcome should be uniform, with an equal (1/6) probability of producing any integer value from 1 to 6. This is illustrated in Fig. 3a. The outcome is determined solely

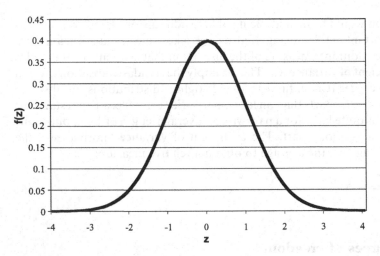

FIGURE 2 Normal distribution.

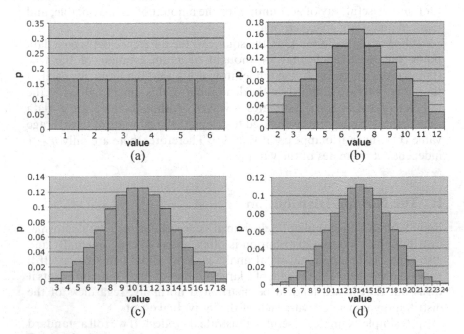

FIGURE 3 Distribution of values obtained from rolling fair six-sided dice. (a) one die, (b) two dice, (c) three dice, (d) four dice.

by the single, random event and so reflects the frequency distribution of that event. If we roll two dice (Fig. 3b), the outcome is associated with two random events. Each is important and the distribution has a distinctive shape reflecting this. With three dice (Fig. 3c), the importance of each single, random event is reduced and the curve begins to assume a somewhat rounded shape. By the time we reach four dice (Fig. 3d) the curve already begins to have a quite bell-shaped, normal-distribution-like appearance.

This property turns out to be important in many applications. For example, observation of a nonnormal distribution in a manufacturing process is an indication that the process may be strongly influenced by a few uncontrolled parameters. Elimination of these should result in much less variability. Similarly, in a well-controlled experiment, once the main potential sources of error are eliminated, the remaining error is often the sum of many small errors from different sources. It is common, therefore, to assume that experimental error produces a normal distribution about the average as a basis of comparison in determining whether an observed effect is statistically significant.

If the average and standard deviation of a distribution of data points are known, and the distribution is assumed to be normal, it is possible to estimate the probability that a measurement that is part of this distribution will be within certain limits. Standardized tables for this are available in most statistics textbooks. For example, the probability of the measurement being within limits of $\pm 1\sigma$ of the average is about 68.3%. Similarly, the probabilities of being within $\pm 2\sigma$ and $\pm 3\sigma$ are about 95.4% and 99.7%, respectively.

In addition, these percentages can also be used to judge whether a measurement differs significantly from that expected from the distribution. For example, there is only a 0.3% (100% − 99.7%) chance that random error would cause an individual measurement to be more than 3σ from the mean. If we take a measurement and observe that its value is outside of this range, we would probably conclude that the value differs significantly from that expected based on random (experimental) error. If the measured value were only $\pm 2\sigma$ from the average (100% − 95.4% = 4.6% probability) we might still conclude that it was significantly different, although with somewhat less confidence.

G. Effect of Finite Sample Size

The above comparisons with the normal distribution assume that the population variance is precisely known. Generally, however, it is necessary to estimate the population variance, σ^2, from the variance obtained with a finite sample, s^2. The additional uncertainty introduced by the need to make this estimate causes a "spread" in the frequency distribution curve. It becomes more probable that a measurement will be obtained many standard deviations from the average value because of uncertainty in the standard deviation itself.

The effects of finite sample size in estimating the variance can be accounted for by replacing the normal distribution with Student's t-distribution, as illustrated in Fig. 4. There are actually many t-distributions, based on the number of degrees of freedom available for estimating the variance. With only 1 degree of freedom (i.e., one independent estimate of the variance), the estimate is not very good and so the t-distribution curve is very broad. With more degrees of freedom, the estimate improves and the t-distribution is narrower. Figure 4 shows curves for 1 degree of freedom and 10 degrees of freedom along with a normal distribution for comparison. In the limit (that is infinite degrees of freedom) the t-distribution is the same as the normal distribution.

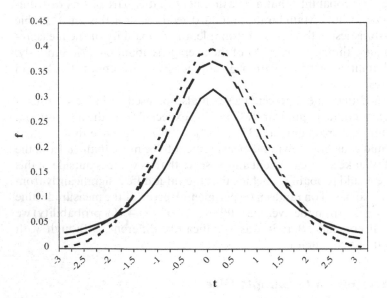

FIGURE 4 Student's t-distribution with 1 dof (broadest curve) and 4 dof (middle curve) compared to normal distribution (narrowest curve).

II. ANOM AND ANOVA

You have already walked through ANOM and ANOVA analyses in Chap. 1, and you will see additional examples later. In this section, you will learn how to calculate the necessary terms for ANOM and ANOVA on an experimental array. Most statistical software packages will perform the algebra involved in these calculations and in practice you will probably end up using one of them to do your actual number crunching. Therefore, although the necessary equations will be provided, the main emphasis is on understanding the underlying logic and assumptions. In doing this, you may find it helpful to draw on the "review" material in the first section for analogies and comparisons.

A. Goals of Analysis

Analysis of the results from a designed experiment is normally directed toward one or more of the following goals:

1. Identifying the "best" levels for the control factors.
2. Ranking of factor and interaction effects in order of importance.
3. Determining the (statistical) significance of effects.
4. Looking for "signs of trouble" in the experiment.

Not every experiment has all of these goals. In particular, judgment of statistical significance is dependent on having an estimate of the error, which is not always available. Of these four goals, ANOM is used mainly for identifying the best levels, although it can also be used to rank effects. ANOVA is useful in achieving the other three goals (2 through 4).

B. Column Identification

Before beginning either ANOM or ANOVA, it is important to confirm the identification of each of the factors and interactions to be examined with a specific column in the experimental array. In the fluid delivery system, for example, Factor A was placed in column 1, Factor B in column 2, etc. (Fig. 8 in Chap. 1). Therefore any results obtained for column 1 are attributed to Factor A in the analysis, etc. (Strategies for assigning factors to particular columns, and determining which columns correspond to which interactions, are presented in Chap. 4. For now assume that, as in the fluid delivery example, this information will be available when you are ready to conduct the ANOM and ANOVA.)

It is also important, to confirm the assignment of particular columns to estimate experimental error. Any interaction terms in these columns are assumed to be zero. In the fluid delivery experiment, for example, all of the higher order interactions (i.e., $A \times B \times C$, $A \times B \times D$, $A \times C \times D$, $B \times C \times D$, $A \times B \times C \times D$) were assumed to be zero, leaving the corresponding columns (5, 6, 7, 9, and 12) for error estimation. To avoid bias, assignment of columns to error estimation should generally be done prior to studying or analyzing the results. (If no columns are assigned to error, ANOVA may still be performed, but judgment of statistical significance may not be possible. See Chap. 8 for a discussion of alternative methods for judging significance.)

C. ANOM Algebra

ANOM addresses the first two possible goals. It can be used to (1) determine which level of each factor is the "best" and (2) provide a relative ranking of their importance. ANOM involves comparing the mean (average) values produced by the different levels of each factor. Specifically, the characteristic responses for all treatment conditions where the factor was at level -1 are averaged and compared to the average obtained with the factor at level $+1$.

The first calculation needed is to determine the overall average, m^*, of the characteristic responses for the experiment:

$$m^* = \frac{\sum_{i=1}^{n} \alpha_i}{n} \tag{5}$$

where α_i is the characteristic response for treatment condition i and n is the total number of characteristic responses. (Note that for the analysis presented here, it is assumed that there is one and only one characteristic response for each treatment condition. Thus, the number of characteristic responses is just equal to the number of treatment conditions. This is not true if replication is used. Replication is discussed in Chap. 8.)

Then for each column in the array, the average of the characteristic response obtained when the column has a level of $+1$, m_{+1}, and when it has a level of -1, m_{-1}, is calculated. Note that although two averages are calculated for each column (m_{+1} and m_{-1}), they must average together to give the overall average for the experiment (m^*), so that there is only one independent term (that is 1 degree of freedom) associated with each (two-level) column. To facilitate comparisons among columns, we can also calculate the difference between these averages:

$$\Delta = m_{+1} - m_{-1} \tag{6}$$

To demonstrate these calculations, consider the calculation for Factor C in the fluid delivery experiment. The necessary data are shown in Fig. 8, Chap. 1. The overall average, m^*, (which is the same for all columns) is given by:

$$m^* = \frac{\sum\limits_{i=1}^{n} \alpha_i}{n}$$

$$= \frac{\alpha_1 + \alpha_2 + \alpha_3 + \alpha_4 + \alpha_5 + \alpha_6 + \alpha_7 + \alpha_8 + \alpha_9 + \alpha_{10} + \alpha_{11} + \alpha_{12} + \alpha_{13} + \alpha_{14} + \alpha_{15} + \alpha_{16}}{16}$$

$$= 5.775625$$

There were eight characteristic responses collected when Factor C had a level of -1; specifically α_1, α_2, α_5, α_6, α_9, α_{10}, α_{13}, and α_{14}. Thus:

$$m_{-1} = \frac{\alpha_1 + \alpha_2 + \alpha_5 + \alpha_6 + \alpha_9 + \alpha_{10} + \alpha_{13} + \alpha_{14}}{8}$$

$$= \frac{7.31 + 2.25 + 6.10 + 5.52 + 2.38 + 1.52 + 0.94 + 3.44}{8}$$

$$= 3.6825$$

Similarly, there were 8 characteristic responses collected when Factor C had a level of $+1$; specifically α_3, α_4, α_7, α_8, α_{11}, α_{12}, α_{15}, and α_{16}, giving:

$$m_{+1} = \frac{\alpha_3 + \alpha_4 + \alpha_7 + \alpha_8 + \alpha_{11} + \alpha_{12} + \alpha_{15} + \alpha_{16}}{8}$$

$$= \frac{8.58 + 8.58 + 10.42 + 9.17 + 5.46 + 7.02 + 7.34 + 6.38}{8}$$

$$= 7.86875$$

Finally, for this column the effect, Δ, is

$$\Delta = m_{+1} - m_{-1} = 7.86875 - 3.6825 = 4.19$$

D. ANOM Comparisons

The absolute size of the effect calculated for each factor and interaction provides a simple basis for comparing their importance. In the fluid delivery example, relative importance of different factors and interactions could be compared in terms of the relative size of Δ in Fig. 8 in Chap. 1. Figure 9 in Chap. 1 showed this same result graphically. The greater, relative importance of Factors A and C was apparent.

E. Determining the "Best" Level for Each Factor

If the experimental goal is to either minimize or maximize the characteristic response, selection of the desired level for each factor based on ANOM is straightforward. The "best" level is simply the level with the smallest (to minimize) or largest (to maximize) average of the characteristic response. For example, to maximize the abrasive concentration in the fluid delivery system, factor levels should be chosen based on which level gives the largest average. As seen in Fig. 8, Chap. 1, this means that level -1 should be used for Factor A (i.e., a convex bottle geometry) since $m_{-1} > m_{+1}$. Graphically this process is conveniently done by selecting the high points of the ANOM graph in Fig. 9, Chap. 1.

If an intermediate level of the characteristic response is desired, selecting levels is more complex. In general, it will be necessary to select a combination of control factor levels, some tending to increase and some to decrease the characteristic response, in order to hit an intermediate target. The "best" response for any individual factor will depend on the levels selected for the other factors. ANOM analysis is useful in establishing the magnitude and direction expected for each factor.

Selection of the characteristic response, including the advantages of a characteristic response optimized at an extreme (maximum or minimum) value, is discussed in detail in Chap. 5.

When interaction terms are involved, selection of the "best" levels is more complicated and is best performed after significance is judged. We will, therefore, return to this topic later in the chapter.

F. ANOVA Concept

ANOVA analysis of an array design is based on breaking the total variance down into distinct components. In particular, the total variance of the characteristic responses is broken down into a contribution from the average and separate contributions from each of the columns in the array. To judge statistical significance, the variance associated with the column associated with a factor or interaction is compared against the average variance associated with columns used for an error estimate. If the factor's or interaction's variance is large in comparison to the error term the factor is judged significant.

G. Preparing the ANOVA Table

The ANOVA table presents the information required for ANOVA in summary form. It can be constructed with five columns, headed Source, SS, dof, MS, and F, respectively. The ANOVA table will have one row for each column used to estimate a factor or interaction effect, one row for the error estimate, and one row for the total. These are labeled in the Source column. (Note that the rows for factors and interactions in the ANOVA table correspond to columns on the experimental array.) Figure 5 shows the table prepared for the fluid delivery example.

H. Calculation for Each Column

For each column of the experimental array we begin by calculating a variance term, expressed as the sum of squares (SS) for the column. For a two-level column the sum of squares is given by:

$$SS \ (2-\text{level column}) = n_{-1}(m_{-1} - m^*)^2 + n_{+1}(m_{+1} - m^*)^2 \quad (7)$$

where n_{-1} and n_{+1} are the number of characteristic responses with a level of -1 or $+1$, respectively, m_{-1} and m_{+1} are the average values of the characteristic response for level -1 or $+1$, respectively, and m^* is the overall average defined by Eq. (5). Note that in calculating the SS, two

Source	SS	DOF	MS	F
A. (Bottle geometry)				
B. (Tube diameter)				
C. (Stir bar)				
D. (Flow rate)				
AxB				
AxC				
AxD				
BxC				
BxD				
CxD				
error estimate				
Total				

FIGURE 5 Table for recording results of ANOVA on fluid delivery system experiment.

values determined from the column were used, m_{+1} and m_{-1}. Only one of these is independent because they must average to give the overall average, m^*. Hence, each two-level column has 1 degree of freedom associated with it.

To demonstrate this calculation, consider again Factor C in the fluid delivery experiment. Values for the overall average, m^*, the average value at level -1 (m_{-1}) and the average value at level $+1$ (m_{+1}) were calculated above. There were eight characteristic responses collected when Factor C had a level of -1; specifically α_1, α_2, α_5, α_6, α_9, α_{10}, α_{13}, and α_{14}. Thus,

$$n_{-1} = 8$$

Similarly, there were eight characteristic responses collected when Factor C had a level of $+1$; specifically α_3, α_4, α_7, α_8, α_{11}, α_{12}, α_{15}, and α_{16}. So,

$$n_{+1} = 8$$

Thus,

$$\text{SS (Factor C)} = n_{-1}(m_{-1} - m^*)^2 + n_{+1}(m_{+1} - m^*)^2$$
$$= 8(3.68 - 5.776)^2 + 8(7.87 - 5.776)^2 = 70$$

I. Calculation for the Total

The total degrees of freedom and total SS entered in the ANOVA table will be the totals without the contribution of the overall average. Because 1 degree of freedom is used in the calculation of the overall average, the remaining degrees of freedom for entry into the total row in the ANOVA table are

$$(\text{Total}) \, \text{dof} = n - 1 \tag{8}$$

The total SS for entry into the ANOVA table is given by

$$(\text{Total}) \, \text{SS} = \sum_{i=1}^{n}(\alpha_i - m^*)^2 \tag{9}$$

J. Filling in the ANOVA Table

For each of the factors and interactions the corresponding SS and dof values are entered into the corresponding rows in the ANOVA table. The

SS and dof for error are obtained by adding the values for all of the columns assigned to error.

The mean square (MS) for the factors, interactions, and error is obtained by dividing the sum of squares by the dof value in each row. To demonstrate, we will calculate the MS values for Factor C and for the error estimate:

$$MS \ (Factor \ C) = \frac{SS \ (Factor \ C)}{dof(Factor \ C)} = \frac{70}{1} = 70$$

$$MS \ (error) = \frac{SS \ (error)}{dof(error)} = \frac{10.64}{5} = 2.1$$

For each of the factors and interactions, the F ratio is calculated by dividing the corresponding MS value by the MS for error. Thus for Factor C:

$$F \ (Factor \ C) = \frac{MS \ (Factor \ C)}{MS \ (error)} = \frac{70}{2.1} = 33$$

Finally the total SS and dof [from Eqs. (8) and (9)] are entered.

For convenience, Fig. 6 (identical with Fig. 10 in Chap. 1) shows the completed ANOVA table for the fluid delivery example.

Source	SS	DOF	MS	F
A. (Bottle geometry)	34.37	1	34.37	**16.1**
B. (Tube diameter)	2.41	1	2.41	1.1
C. (Stir bar)	70.10	1	70.10	**32.9**
D. (Flow rate)	1.35	1	1.35	0.6
AxB	0.48	1	0.48	0.2
AxC	0.35	1	0.35	0.2
AxD	5.21	1	5.21	2.4
BxC	0.08	1	0.08	0.04
BxD	1.04	1	1.04	0.5
CxD	0.70	1	0.70	0.3
error estimate	10.64	5	2.13	-
Total	126.83	15	-	-

FIGURE 6 Completed ANOVA table for fluid delivery system experiment.

K. Check on the Totals

Assuming that all of the dof and SS are accounted for by the columns in the matrix, the "total" values included in the ANOVA table should just be equal to the sum of the contributions from all of the rows in the table. This provides a check on the matrix and calculations. For the fluid delivery example, summing up the dof for all of the factors, interactions, and the error in Fig. 6 gives a total of 15, which matches that obtained from Eq. (8) above. Similarly, summing the SS for these same rows gives a total of 126.73. Again, this agrees, within rounding error, with the total calculated from Eq. (9) and shown in the bottom row of the table. (Note that for some of the more complex arrays developed later in this book, not all of the dof and SS are represented in column form. In this case a difference between the total value and the sum from the individual columns is expected and provides a way to calculate the noncolumn values. This is discussed in Chap. 7.)

L. Ranking and Significance

As with ANOM, the results of ANOVA can be used to rank the relative size/importance of the various factors and interactions. In ANOVA, this is conveniently done based on the relative size of the SS or the F ratio calculated for each effect. Of course, such a ranking is meaningful only if the measured effects are the result of the specified factor or interaction and not of experimental error. This is assessed by judging the statistical significance.

In ANOVA, statistical significance is judged based on the F ratio. A very large value of F corresponds to a variance much larger than that expected from error and, therefore, "probably real." Whether the experimental F is "large enough" depends on the confidence level desired (expressed in terms of P), the dof in the factor/interaction, and the dof used for the error estimate. These three parameters are combined to give a critical F ratio, F_{cr}. If the measured F is larger than F_{cr}, the factor or interaction is judged statistically significant. Rigorous application of this test is dependent on assumptions about the data set that in practice are often not fully justified (nor tested). Hence, the analysis should be considered only an approximation for the applications described here. It nevertheless provides a reasonable, objective, criteria for judging how large an F value is required for a factor or interaction to be judged significant.

A P of 0.90 means there is an approximately 1 in 10 chance $(1.0 - P)$ that an F value as large as F_{cr} would occur randomly (i.e., due to experimental error). A P of 0.99 means the chance is approximately 1 in 100. More dof for error gives a better estimate of the true error variance and therefore lowers the required value of F_{cr} (analogous to the narrowing of the t-distribution with increasing dof). Values of F_{cr} are often specified using a nomenclature of

$$F_{cr} = F_{1-P,v_1,v_2}$$

where v_1 is the degrees of freedom for the factor or interaction and v_2 is the degrees of freedom for the error.

Tables giving F_{cr} as a function of these three parameters are provided in Appendix D. For example, for a P of 0.90, with 1 dof for the factor/ interaction and 5 dof for error, $F_{cr} = F_{0.1, 1,5} = 4.06$. Applying this value to the fluid delivery ANOVA table (Fig. 6), it is seen that the F values for Factor A ($F = 16.1$) and C ($F = 32.9$). exceed the critical value and would be judged significant.

Using a higher confidence limit of $P = 0.99$, under these same conditions, one obtains $F_{cr} = F_{0.01, 1,5} = 16.26$. Factor C would still be judged significant ($F > F_{cr}$), although Factor A would be considered marginal ($F < F_{cr}$, but so close that in practice we would probably still include it.)

None of the other factors and effects have a large enough F ratio. In particular, note that A $\times$ D has an F ratio of 2.4 (i.e., its mean square is more than twice the value for the error estimate) but is not judged significant. It is important to understand that just because a factor or interaction is not judged significant does not mean that it has no effect. It merely means that the observed effect was not large enough to be satisfactorily distinguished against the background produced by the error.

M. Signs of Trouble

Before concluding an ANOVA, it is important to also look at the results with an eye toward detecting signs of possible trouble in the analysis.

(1) The first, and most obvious, trouble sign is the lack of any large (significant) F ratios. This is an indication that the effects of the factors being studied cannot be resolved against the error background. Three probable causes may be distinguished. First, it is possible that the factor effects are genuinely small. Reexamination of the system to identify other possible control factors should be considered. A second possibility is that the experimental error is large. The experimental setup should be reexamined with a view to identifying possible sources of error, such as

uncontrolled variables, inadequate instrumentation, poor measuring technique, etc. A third possibility is that the real (experimental) error has been overestimated. Error estimates are frequently made by assuming that certain columns are "empty," in other words, by assuming that all of the interaction terms in the column are zero. In the fluid delivery study, e.g., error was estimated by assuming that all interactions involving three or four factors were zero. If one or more of these terms is actually large, the error estimate may reflect this "modeling error" (error in our assumptions) rather than the experimental error. The assumptions used in setting up and analyzing the experiment may need to be reconsidered. In addition, it may be valuable to also analyze the results using the normal-probability-plot technique, which does not require up-front assumptions about which columns are "empty." (Use of normal probability plots will be discussed in Chap. 8.)

(2)   A second sign of potential trouble occurs when a large percentage of the total SS in the experiment is attributed to error. This is a subjective judgment and very dependent on the system and its maturity. During preliminary research, "unknown" (modeling or experimental) errors may well constitute a relatively large fraction of the total variance in the system. However, for a more "mature" product or process, having a large fraction of the variance unaccounted for is a serious concern. The system should be reexamined for possible sources of experimental or modeling error as described above.

(3)   A third potential trouble indicator is the occurrence of a large interaction term involving factors that are, by themselves, small. Although this is not impossible, more commonly large interactions involve factors that also have large independent effects. The confounding pattern of the experimental design should be reexamined to look for possible alternative interpretations. For example, is it possible that confounding with another interaction, originally assumed to be zero, is responsible for the observed effect? (An example is given in Section IV of Chapter 8.)

N. Determining "Best" Levels with Interactions

The level of an interaction term is determined by the levels set for the factors involved and cannot be set independently. Therefore, when an interaction is judged significant it may be necessary to consider it together with the factors in determining best settings. In practice, it is relatively rare for interaction effects to strongly influence best factor settings. This is one reason screening-type experiments (that focus only on isolating factor effects) are often effective. However, a brief discussion is presented here for completeness.

One advantage of using the -1, $+1$ nomenclature to indicate the levels of two-level factors is that it permits the level of an interaction to be easily determined from the level of the factors. Specifically, the level of an interaction is equal to the product of the levels of the factors involved. To illustrate, consider the A × B interaction. A × B has a level of $+1$ when Factors A and B are either both at $+1$ or both at -1 ($+1 \times +1 = +1$, $-1 \times -1 = +1$). A × B has a level of -1 when A is at -1 and B at $+1$, or when A is at $+1$ and B at -1 ($-1 \times +1 = -1$, $+1 \times -1 = -1$). [The arrays and other information in Appendix A are written consistent with this convention, which is widely, but not universally, used. You may confirm the applicability of this method for any of the arrays in Appendix A by checking the levels of any two columns against the column corresponding to their interaction]

If the best level for an interaction corresponds to the best level of the factors involved, we say the interaction is synergistic. In this case, the interaction term increases the improvement expected from the factors but does not alter determination of the best factor levels.

If the best level for an interaction does not correspond to the best level of the factors involved, we say it is antisynergistic. It works against the improvement expected from the factor effects alone. Whether the best settings for the factors are altered depends on the relative magnitude of the factor and interaction effects. Two cases may be distinguished based on whether one or both of the factors is judged significant. If the interaction and only one factor are significant, the level of the significant factor should be set to its best setting first. The level of the second (not significant) factor is then selected to give the desired level for the interaction.

If the interaction and both factors are significant, the various possible factor combinations may be compared to determine the best set. This may also be necessary when there are multiple interactions involving several factors. Such comparisons can be conveniently accomplished by trying the various possible combinations in a model such as that described in the next subsection. (Interaction plots, described in Chap. 4, may also be helpful.)

O. Model

Using the results from the ANOM and ANOVA analyses, one may develop a simple model to provide a first estimate of the characteristic response that would be obtained under specific factor settings. Comparison of the predictions of this model against results of subsequent experiments also provides an additional check on the analysis.

The model is constructed by adding to the overall average (m^*) a term for each significant factor and interaction. The term for each factor/interaction is equal to the difference between the average value of the characteristic response for the chosen level and the overall average. Hence it will be positive when the average response for the level is greater than m^*, and negative when it is less.

To illustrate, assume that Factors A and B and the interaction A × B are all significant. If both A and B are set to level +1 (corresponding to A × B also at +1):

$$\alpha_{pred} = m^* + (m^A_{+1} - m^*) + (m^B_{+1} - m^*) + (m^{A \times B}_{+1} - m^*)$$

where α_{pred} is the characteristic response predicted under the specified settings, m^* is the overall average, m^A_{+1} is the average characteristic response when Factor A is at level +1, m^B_{+1} is the average characteristic response when Factor B is at level +1, and $m^{A \times B}_{+1}$ is the average characteristic response when the interaction A × B is at level +1.

If A is set to level +1 and B to −1 (corresponding to A × B at −1), then

$$\alpha_{pred} = m^* + (m^A_{+1} - m^*) + (m^B_{-1} - m^*) + (m^{A \times B}_{-1} - m^*)$$

Similarly, α_{pred} can be calculated for other factor level combinations.

Finally, consider the application of this approach to the specific example of the fluid delivery study. Only Factors A and C were judged significant, with best levels of −1 and +1, respectively. Thus the predicted response under the optimum conditions is

$$\alpha_{pred} = m^* + (m^A_{-1} - m^*) + (m^C_{+1} - m^*)$$
$$= 5.78 + (7.24 - 5.78) + (7.87 - 5.78) = 9.33$$

which compares reasonably well with the results of the subsequent testing under these conditions (Fig. 11, Chap. 1).

Homework/Discussion Problems

(1) A computer simulation study was run to examine the effects of two factors (A, B) and their interaction (A × B) on part shrinkage in sintered powder metallurgy parts. The 4 TC factorial array shown in Fig. P.1 was

TC	A	B	A × B	Noise conditions		α_i
				1	2	
1	-1	-1	+1	6.5	7.7	
2	-1	+1	-1	10.0	1.0.2	
3	+1	-1	-1	4.0	6.6	
4	+1	+1	+1	6.0	7.0	

FIGURE P. 1 Data for problem 1.

used. Two noise conditions were used, and the response was measured in terms of the percent linear shrinkage. Results are shown in Fig. P.1.

In this problem, assume that the average linear shrinkage is to be used as the characteristic response.

(a) Determine the characteristic response (α = average linear shrinkage) for each treatment condition and enter it into the table above.

(b) Conduct an ANOM analysis of this data, and show the results in tabular and graphical form.

(c) Discuss the relative importance of the two factors and the interaction, as revealed by the ANOM analysis.

(2) An experiment was conducted to evaluate the efficiency of toner use from the toner bottle in a copier. Five control factors (A–E) were tested in an array with eight treatment conditions. "Noise" in the system was believed to consist mostly of the ambient conditions (temperature, humidity) and these were incorporated in the form of three noise conditions. For each combination of treatment and noise conditions the response of the system was measured as the percentage of the initial toner mass that could be extracted (used) before the bottle needed to be replaced.

Figure P.2 summarizes the experiment, including the raw data in the form of the response measured for each combination of treatment and noise conditions. All interactions have been assumed to be zero. Columns containing only interactions are therefore to be used for the error estimate. These are marked with an "e" in the table below. Responses for each set of control factors are summarized below.

For this problem use the average of the percent toner mass used (average of the responses) as the characteristic response.

(a) Calculate the characteristic response for each treatment condition and enter the values into the table.

TC	A	B	C	e	D	E	e	Noise Condition			α_i (%)
								1	2	3	
1	-1	-1	-1	-1	+1	+1	+1	77.9	81.0	84.1	
2	-1	-1	+1	+1	-1	-1	+1	78.4	81.6	85.4	
3	-1	+1	-1	+1	-1	+1	-1	83.0	83.2	88.5	
4	-1	+1	+1	-1	+1	-1	-1	82.0	81.9	84.8	
5	+1	-1	-1	+1	+1	-1	-1	81.2	83.2	85.7	
6	+1	-1	+1	-1	-1	+1	-1	86.6	88.2	89.8	
7	+1	+1	-1	-1	-1	-1	+1	88.9	88.5	89.3	
8	+1	+1	+1	+1	+1	+1	+1	84.5	89.8	89.7	

FIGURE P. 2   Data for problem 2.

(b) Conduct an ANOM analysis. Present your results in both tabular and graphical form. Determine the "effect" for each factor and identify the "best setting."

(c) Conduct an ANOVA analysis and prepare an ANOVA table in the standard form. Judge the significance of the five factors and rank them in order of importance. (Use $P = 90\%$.)

(d) Would your answer to part (c) differ if you had used a different value of P? Discuss.

(e) Are there any signs of trouble in the ANOVA table prepared for (c)? Discuss.

(f) Predict the average toner mass extracted using the best settings (i.e., levels) of all of the significant control factors.

(3) Analysis of a 16 TC factorial array experiment with four factors placed to give a full factorial design produced the sum of squares and degrees of freedom data shown in Fig. P.3 for each of the factors and interactions.

(a) Construct an ANOVA table for this experiment, using higher order (three- and four-factor) interactions to estimate error.

(b) Which effects are judged to be significant? (Be sure to show the necessary "critical" F value.). Use a 95% confidence.

(c) Are there any "signs of trouble"?

(4) An experiment was conducted to evaluate the curing process for a polymer. The initial part length was 30 mm and the desired (cured) part length was 20 mm. Five control factors were tested using an experimental design with eight columns and the factors placed in the first five columns, as shown in Fig. P.4. (This particular design is based on the 8 TC factorial

Source	SS	DOF
A	9312.25	1
B	0.25	1
C	2.25	1
D	1225	1
AB	9	1
AC	0	1
AD	182.25	1
BC	4	1
BD	0.25	1
CD	12.25	1
ABC	0.25	1
ABD	4	1
ACD	16	1
BCD	0	1
ABCD	0.25	1

FIGURE P. 3   Data for problem 3.

TC	A	B	C	D	E	e	e	α (mm)
1	-1	-1	-1	-1	+1	+1	+1	17
2	-1	-1	+1	+1	-1	-1	+1	18
3	-1	+1	-1	+1	-1	+1	-1	20
4	-1	+1	+1	-1	+1	-1	-1	21
5	+1	-1	-1	+1	+1	-1	-1	23
6	+1	-1	+1	-1	-1	+1	-1	29
7	+1	+1	-1	-1	-1	-1	+1	28
8	+1	+1	+1	+1	+1	+1	+1	25

FIGURE P. 4   Data for problem 4.

array, which is discussed in Chap. 4.) All interaction terms have been assumed equal to zero. The length of the part was used as the response.

"Noise" in the system was believed to consist of the ambient conditions (temperature, humidity) and the "age" of the raw materials used in the process. These were incorporated in the form of 10 noise conditions. For convenience, the responses obtained for the 10 noise conditions have been summarized in Fig. P.4 in terms of the average length after curing (α_i) which should be used as the characteristic response for each treatment condition. Estimate error using the columns that contain only interaction terms (labeled "e" in Fig. P.4).

(a) Conduct an ANOM analysis. Present your results in both tabular and graphical form. Determine the "effect" for each factor and identify which level for each factor tends to increase the cured length.

(b) Conduct an ANOVA analysis and prepare an ANOVA table in the standard form. Judge the significance of the five factors and rank them in order of importance. (Use $P = 90\%$.)

(c) Would your answer to part (b) differ if you had used a different value of P? Discuss.

(d) Are there any signs of trouble in the ANOVA table prepared for (b)? Discuss.

(e) Predict the average length of cured parts expected if all factors were set to their -1 level. Be sure to include only factors judged significant in your model.

4

Array Design
(Two-Level Factors)

Overview

In this chapter, you will learn how to design an experimental array. The chapter is divided into three sections. In Sec. I, the basic concepts are presented. In Sec. II, the role of interactions in array design is described. In Sec. III, fundamental trade-offs are discussed, a menu of array design strategies is provided, and some examples drawn from the technical literature are presented to illustrate the strategies.

This chapter deals with experimental designs involving factors with two levels. Extension of these principles to factors with three levels and modifications to the standard array designs are described in Chap. 7.

I. BASIC CONCEPTS

A. Array Design

To conduct a designed experiment, each control factor's level (-1 or $+1$) must be specified for each treatment condition. This is the basic goal of array design. We define an "array" as a specific set of columns, containing various arrangements of the possible factor levels (-1, $+1$). You have

55

already seen one array, with 16 rows and 15 columns, in Fig. 3 of Chap. 1. Array design involves two important steps:

1. Selection of a specific array
2. Assignment of factors to specific columns within the array

Appendices A (Two-Level Factorial Arrays) and C (Screening Arrays) provide a library of useful and popular arrays for two-level factors along with various additional information to help in array design. (Note that the 18 TC Screening array in Appendix C is not in this category. It should be set aside for the moment. Its application is discussed and explained in Chap. 7.)

To illustrate the use of these appendices, consider the first (and simplest) of the arrays presented, the 4 TC Factorial Array, and assume that we have two factors to test. Figure 1 shows the first portion of the information presented in Appendix A. In the upper right-hand corner is the array structure, consisting of three columns. Selecting this array defines the "columns" we will use in conducting and analyzing the experiment. The bottom portion of this table provides recommendations about where to place the factors. With two factors, recommended placement is in columns 1 and 2 of the design. Information below (in the notes and specifications sections) provides additional information about the design. In

				Array		
			TC	Columns		
				1	2	3
			1	-1	-1	+1
			2	-1	+1	-1
			3	+1	-1	-1
			4	+1	+1	+1
# of factors	Resolution					
2	Full	Recommended factor placement		A	B	
3	III	Recommended factor placement		A	B	C

FIGURE 1 Four TC Factorial Array from Appendix A.

particular, it can be determined that with this factor placement, the A × B interaction is located in column 3 of the array. Figure 2 shows the completed array design for this experiment.

A second example of the array design process was provided with the fluid delivery system example in Chap. 1. In particular, the 16 TC Factorial Array from Appendix A was selected. The four factors were assigned to columns 1–4 as recommended. Location of the two-factor interactions was determined using the interaction table for this design from Appendix A. Higher-order interactions were also located but marked with an "e" for use in the error estimate. This produced the completed array design as shown in Fig. 3 of Chap. 1.

You will see many more examples of this array design process as you go through this chapter and book. To make this process more meaningful, however, it is necessary to first understand more about the structure of the arrays, especially the role of interactions.

B. Equivalence of Arrays

The array designs given in Appendices A and C can be generated from scratch, producing the necessary "columns" for each factor based on first principles. This is somewhat cumbersome, but executed properly will automatically "place" factors in the correct columns to produce designs with good overall configurations, i.e., the highest possible "resolution" as described later in this chapter. Many commercial software programs reduce the work involved in this process, generating designs with the highest resolution for the specified array and number of factors. In contrast, Taguchi eliminated the need to generate arrays by supplying a set of standard columns for each of his arrays, allowing the designer to immediately place factors in columns. This system is easy to learn and does

TC	\multicolumn{3}{c}{Columns}		
	A	B	AxB
1	-1	-1	+1
2	-1	+1	-1
3	+1	-1	-1
4	+1	+1	+1

Array

FIGURE 2 Four TC Factorial Array with column assignments for full factorial design.

not automatically place factors in specific columns. This gives flexibility to meet unusual conditions, but does not emphasize the generation of high-resolution designs.

The system used here, and by some other authors as well, combines some features from both of these approaches. For each array, a standard set of columns is provided, along with a list of specific factor placements that produce high-resolution designs. Producing a high-resolution design is straightforward and the designer also has the option of making alternative placements if desired or necessary.

It is important to understand that, regardless of how they are generated or displayed, the underlying array structures are a consequence of mathematical and statistical principles. Different approaches therefore tend to converge on the same basic structures. For example, Fig. 3 shows three matching designs: (1) the 4 TC factorial array from Appendix A, (2) a factorial array generated "from scratch," and (3) Taguchi's $L4(2^3)$ array. In this case, the first two arrays match exactly. The Taguchi array is superficially different; however, this is mainly a function of the arbitrary coding of levels (-1, $+1$ vs. 1, 2). In addition, the level coding (low vs. high) in the third column of the Taguchi array is reversed from the other arrays. This does not alter the underlying symmetry and functioning of the array. These three arrays are functionally equivalent. Note also that in this example, the columns are written in the same order in all three arrays (i.e., column 1 = column 1 = column 1, etc.). This is not always the case, but, again, the order in which the columns are written is a superficial difference that does not affect the underlying function. (However, it does mean that care must be taken to use a consistent set of information. Recommended factor placements, information on the location of interactions, etc. will be altered if the order of the columns is different.)

TC	Columns			TC	Columns			TC	Columns		
	A	B	AxB		A	B	AxB		A	B	AxB
1	-1	-1	+1	1	-1	-1	+1	1	1	1	1
2	-1	+1	-1	2	-1	+1	-1	2	1	2	2
3	+1	-1	-1	3	+1	-1	-1	3	2	1	2
4	+1	+1	+1	4	+1	+1	+1	4	2	2	1

(a)   (b)   (c)

FIGURE 3 Equivalent experimental designs. (a) From Appendix A, (b) from "scratch," (c) Taguchi $L4(2^3)$ orthogonal array.

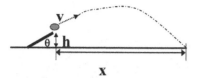

FIGURE 4 Catapult.

The fact that the underlying arrays are functionally equivalent does not mean that all design approaches will produce identical designs. It merely means that all designers are working within the same basic constraints. Differences between different designers and design philosophies are reflected in both array selection and assignment of factors to specific columns within the array. The specific examples presented in this book have been drawn from the technical literature in a wide variety of fields. In some cases, the designs were generated from "scratch," in others, by using commercial software, and, in still others, by using Taguchi's orthogonal arrays. Regardless of the origin, all are described and discussed here using a consistent nomenclature and terminology. Removing the superficial differences allows the fundamental differences in design philosophy to be presented more clearly and consistently.

C. Catapult Example

To provide an example for explanatory purposes, it is useful to "flesh out" the array design shown in Fig. 2 with some additional details. Assume that we wish to study the distance that a catapult will propel differently shaped projectiles (Fig. 4). Launch height (h) and launch angle (θ) are chosen as control factors for study, with the two levels selected as shown in Fig. 5. It will be assumed that the launch velocity is held constant. To complete the example, we assume that for each treatment conditions, the distance traveled (response) is measured for different projectile shapes (noise

Control Factor	Level	
	-1	+1
A. height, h_0 (cm)	15	25
B. launch angle, θ	45°	60°

FIGURE 5 Control factor levels for catapult example.

TC	Column (Factor/Interaction)			α_i
	1 (A)	2 (B)	3 (AxB)	
1	-1	-1	+1	α_1
2	-1	+1	-1	α_2
3	+1	-1	-1	α_3
4	+1	+1	+1	α_4

FIGURE 6   Array design for catapult experiment.

conditions) and used to calculate an average distance, α_i (the characteristic response). For simplicity, only α_i is shown in the final design in Fig. 6.

D.   Comparison with One-At-A-Time Experiments

One of the first lessons most people are taught about experimentation is to change only one variable (factor) at a time. If you change several, you will be unable to tell which of them is responsible for the differences you observe. This is good advice, but does not apply to experiments using properly constructed array designs. This is because, although multiple factors are changed within the experiment, they are changed in a carefully controlled pattern. In fact, the need to have such a pattern is what gives the arrays their basic structure.

Figure 7 shows the one-at-a-time approach, applied to the catapult experiment described above. In this case, there are two sets of experiments; in the first, the launch height (factor B) is fixed (at level −1) and the angle (factor A) is changed (between level −1 and level +1). In the second, the angle is fixed and height is varied.

First, let us consider the size of the experiment. Although there are nominally four treatment conditions in Fig. 7, two of these have identical

TC	Column (Factor)		α_i
	A	B	
1'	-1	-1	α_1'
2'	-1	+1	α_2'

3'	-1	-1	α_3'
4'	+1	-1	α_4'

FIGURE 7   "One-at-a-time" design for catapult experiment.

control factor settings; that is, TC 1′ and TC 3′ both correspond to A and B both at level −1. It would therefore only be necessary to run three different sets of control factors (i.e., three unique treatment conditions), in contrast to the four needed for the array design (Fig. 5). As will be discussed later, there is some flexibility in the array design. It is possible to increase the number of factors (or decrease the array size) by accepting confounding of factor and interaction effects. Nevertheless, the basic conclusion is correct: one-at-a-time designs can often be made smaller. This is particularly true if the number of levels per factor is large. Array experiments are most competitive when the number of levels per factor is small (two or three). One-at-a-time experiments become increasingly attractive if it is necessary to include many levels for the factors.

Second, consider the role of experimental error. For the one-at-a-time approach, the effect of A (i.e., $\Delta = m_{+1} - m_{-1}$) is given by the difference of two numbers, $\alpha_2' - \alpha_1'$. In contrast, the array design determines the effect based on four numbers, $(\alpha_3 + \alpha_4) - (\alpha_1 + \alpha_2)$. In consequence, the array design is less subject to random experimental error in any of the individual characteristic responses. To duplicate this effect in the one-at-a-time experiment, the experiment could be replicated (run multiple times) at the expense of increasing the required effort. The "pseudo-replication" available in array designs is one of their main advantages and scales with the array size. For example, for the fluid delivery experiment (16 TC Factorial Array design) described in Chap. 1, each effect was based on values for 16 characteristic responses.

Finally, note that for the one-at-a-time approach, the effect of each factor is evaluated at only one value of the other factor(s). In contrast, for the array design, the effect of one factor is averaged over different values (levels) for the other factor(s). In an industrial design setting, this can be an important advantage since it avoids the risk of identifying the "best" settings (levels) for each factor based on specific settings of all the other factors, which may or may not be those selected for the final design.

II. INTERACTIONS

A. Definition of an Interaction

Interactions are among the most frequently misunderstood aspects of array design. Simply stated, an interaction occurs when the difference measured between the levels for one factor depends on the level of another factor. To illustrate, consider measuring plant growth as a function of sunlight and water, with the −1 level of both factors indicating "none" (no sun or no water). Since growth requires both sunlight and water, the effectiveness of adding water is strongly dependent on whether or not there is sunlight and vice versa. In an array design, this interdependence will appear as an interaction effect.

Interactions can occur between all possible combinations of the factors in an experiment, including those involving more than two factors. For example, with four factors (A, B, C, and D), there are 11 interaction terms (A × B, A × C, A × D, B × C, B × D, C × D, A × B × C, A × B × D, A × C × D, B × C × D, and A × B × C × D). However, attention is often focused on two-factor interactions, which are generally considered more likely to be large (significant).

B. Interaction Graph

A useful tool to visualize and work with interactions is the interaction graph. Figure 8 shows a schematic interaction graph for the plant growth experiment described above. Average plant growth rate (characteristic response) is plotted vs. the presence of water (control factor) as in a normal ANOM plot. For the interaction graph, however, the data are split to show separate averages for each of the levels of the other control factor, in this case, the sunlight. The existence of an interaction is seen in the difference in slope between the two lines. In this particular case, it is easy to see that the setting of the second control factor (sunlight) has a large effect on what happens when the level of the first control factor (water) is changed.

Figure 9 shows a real interaction graph based on a recent experimental study of an adhesion tester [2]. In this work, the effects of five

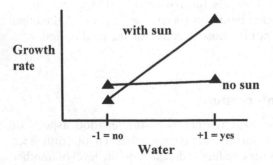

FIGURE 8 Schematic interaction graph for plant growth.

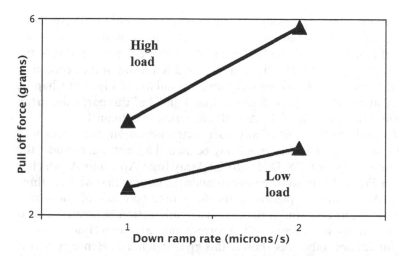

FIGURE 9 Interaction graph for load vs. ramp rate effect on pull-off force. (Adapted from Ref. 2.)

control and noise factors and their interactions on adhesion was studied. This graph was constructed by calculating four average characteristic response values, representing the four possible combinations of levels for two of the factors (applied load and removal rate). Specifically, an average was calculated for combinations of low load and low rate, low load and high rate, high load and low rate, and high rate and high load. These four averages were then plotted as points to give the interaction graph in Fig. 9. The difference in the slopes is subtler than in Fig. 8, but shows a real effect of the load/rate interaction on pull-off force.

Note that in an interaction graph, it is the difference in the *slope* that denotes the presence of the interaction and not the difference in "height." The height difference reflects the effect of the factor by itself. Whether this difference is significant can be judged from the ANOVA result for the interaction (e.g., A × B) term. Finally, note that if A interacts with B, B necessarily has a corresponding interaction with A (A × B = B × A).

C. Location of Interaction Terms (Factorial Arrays—Appendix A)

In an array design, interactions may be treated somewhat like factors with levels that can be used to estimate their effect (Δ). In contrast to factors, however, interactions cannot be assigned independently to columns. Instead, the location of an interaction is determined by the array layout and by the placement of the factors involved. For example, in Fig. 6, the location of the A × B interaction in column 3 is a result of the decision to assign factors A and B to columns 1 and 2. Similarly in Fig. 3 of Chap. 1, assignment of factors A and B to columns 1 and 2 of this particular array determines the location of the A × B interaction in column 15.

To find the location of any particular interaction, the interaction table for the design (Appendix A) may be used. To illustrate, consider the interaction table for the 8 TC Factorial Array from Appendix A, which is shown in Fig. 10 (with some emphasis added). Assume that we have three factors (A, B, and C) placed in the first three columns of the array, respectively. The location of the two-factor interaction between A and B is found from the intersection of the corresponding column (1) and row (2) in the interaction table. As shown, this gives column 7. Hence A × B is located in column 7.

To find higher-order terms, this process can be continued. For example, since A × B is in column 7 and C in column 3, the location of A × B × C can also be found from the intersection of the corresponding column (3) and row (7) in the table, which gives column 4. Hence A × B × C is located in column 4.

Thus factor assignment determines interaction location. As you will see below, some combinations of factor and interaction locations are more favorable than others. This is why factor assignment is an important step in array design. Appendix A includes tables and other information to

		Column						
		1	2	3	4	5	6	7
C	1	-						
o	2	7	-					
l	3	6	5	-				
u	4	5	6	7	-			
m	5	4	3	2	1	-		
n	6	3	4	1	2	7	-	
	7	2	1	4	3	6	5	-

FIGURE 10 Interaction table for 8 TC Factorial Array, illustrating determination of interaction location.

determine the location of interactions based on factor assignment. Application examples are provided later in this chapter. Note that this description applies to the factorial array designs in Appendix A. The screening arrays in Appendix C follow somewhat different rules and will be described later in this section.

D. Confounding

Because of the large number of possibilities, measuring all interaction terms can consume disproportionate amounts of experimental effort. Moreover, in many practical applications, interactions are found to be small compared to the main (factor) effects. For example, in the fluid delivery study, none of the two-factor interactions was determined to be statistically significant. As a result, there is a strong incentive in experimental design to assume up front that interaction terms will be small. This allows effort to be focused on measurement of factor effects, but introduces uncertainty.

Consider, for example, the third column in the array design shown in Fig. 6. If during design of the array the assumption is made that the interaction effect (A × B) will be zero, the third column can be used for other purposes. For example, it could be used as an error estimate to help judge the significance of the results. Alternatively, we could place an additional factor ("C") into this column and measure the effects of three factors instead of two for the same total experimental effort (i.e., number of treatment conditions). The price paid for this is the introduction of additional uncertainty into the analysis. When we change the level of factor C in the experiment, we are also changing the level of A × B. As a result, if the effect of A × B is not zero, it will be superimposed on that of C. There is no way to determine from the experimental results whether any effect we calculate from column 3 is actually caused by factor C, or by the A × B, or by some combination of the two. This situation is summarized by saying that C and A × B are confounded with each other. When two or more factors or interactions are confounded, we cannot tell from the experimental results which (or which combination) is the true cause of the effect we calculate for the column.

Placing C in column 3 also creates additional confounding due to its interaction with other factors. Figure 11 shows the completed pattern. In this case, each factor is confounded with one interaction (A with B × C, B with A × C, and C with A × B). To determine the factor effects with this design, it would be necessary to assume that the effects of all three of these

TC	Column (Factor/Interaction)			α_i
	1 (A, BxC)	2 (B, AxC)	3 (C, AxB)	
1	-1	-1	+1	α_1
2	-1	+1	-1	α_2
3	+1	-1	-1	α_3
4	+1	+1	+1	α_4

FIGURE 11   Four TC Factorial Array with three factors, illustrating confounding of factors and interactions.

interactions were zero. With larger designs and more factors, much heavier and more complex confounding can result.

E.   What Is "Wrong" with Interactions?

From one perspective, interactions are simply part of the mathematics of array design and analysis. Their magnitude can be calculated and the results can be incorporated into final decision making. From a practical standpoint, however, the presence of strong interactions among the control factors is generally undesirable. First, since the number of possible interactions grows very rapidly with the number of factors, considerable experimental effort is needed to measure each of them. In most cases, we would rather expend this effort tracking the effects of the factors than looking for possible interaction terms. The design in Fig. 11, for example, allows us to test the effects of three factors with the same number of treatment conditions used in Fig. 6 to test only two factors.

Additionally, the presence of large interaction terms complicates the information from the experiment, making it more difficult to convey and implement during subsequent design and manufacturing steps. If, for example, there is a strong interaction between two factors, any subsequent change in one of them (possibly due to manufacturability issues) may require changes in the other. If multiple interactions are involved, a cascade of changes can result.

In dealing with interactions in experimental design, a two-tiered approach is useful:

1.   Existing understanding of the system under study can be used to reduce the likely strength of interactions appearing in the analysis.

2. Since the occurrence of strong interactions cannot be completely ruled out, array design is used to control the pattern of confounding between factors and interactions.

F. Reducing the Strength of Interactions

At first glance, it may appear that interactions are a result of the basic physics of the system under study and therefore not subject to control. However, the apparent complexity of a problem, which includes the presence or absence of strong interactions, is also a function of the conceptual framework used to analyze and describe the results. As an example, consider the motion of a satellite in a circular (nongeosynchronous) orbit about the earth. Expressed in terms of Cartesian (X–Y) coordinates relative to an arbitrary point on the earth's surface, a relatively complex set of equations is necessary. However, expressed in terms of a polar (ρ–θ) coordinate system about the earth's center, the mathematical description is greatly simplified. Similarly, in experimental design, the complexity of the final results (including the relative strength of interactions) is dependent on design choices, such as the "coordinate system" imposed by selection of the control factors.

To illustrate, consider applying this idea to the catapult example described previously. The launch height and launch angle (h_0 and θ) can be used as control factors as illustrated in Figs. 12a and 13a. However, in actually setting up the catapult, the launch height and angle are determined by specifying the heights (h_1 and h_2) of the two ends of the launching ramp as illustrated in Fig. 12b. Selecting h_1 and h_2 as control factors produces the

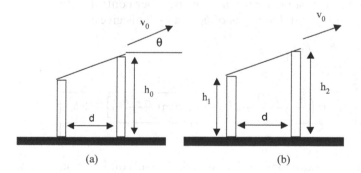

(a)   (b)

FIGURE 12 Schematic of catapult (a) with h_0 and θ as control factors; (b) with h_1 and h_2 as control factors.

Factor/level	1	2
A) h_0 (cm)	15	25
B) θ (degrees)	45	60

(a)

Factor/level	1	2
A) h1 (cm)	0	10
B) h2 (cm)	15	25

(b)

FIGURE 13   Control factor levels for catapult (a) with h_0 and θ as control factors; (b) with h_1 and h_2 as control factors.

alternative array design shown in Fig. 13b. Which design will produce a simpler result?

From dynamics, the distance a projectile will travel from its launch point to impact, x, is given by the equation:

$$x = v_0 \cos \theta \left[\frac{v_0 \sin \theta + \sqrt{v_0^2 \sin^2 \theta + 2gh_0}}{g} \right] \tag{1}$$

where g is the acceleration due to gravity and v_0 is the launch velocity. Notice that the fundamental quantities involved in describing the physics of this problem are v_0 (held constant), h_0, and θ, matching the first set of possible control factors (Fig. 13a).

We can obtain a similar equation for the other control factor set by substituting in for h_2 and θ in terms of h_1 and h_2. This gives:

$$x = v_0 \cos \left[\arctan \left(\frac{h_2 - h_1}{d} \right) \right]$$

$$\times \left\{ \frac{v_0 \sin \left[\arctan \left(\frac{h_2 - h_1}{d} \right) \right] + \sqrt{v_0^2 \sin^2 \left[\arctan \left(\frac{h_2 - h_1}{d} \right) \right] + 2gh_2}}{g} \right\} \tag{2}$$

where d is the (fixed) horizontal separation of the ends of the launch ramp. Although the designs in Fig. 13a and b are very similar in appearance, they lead to solutions of very different complexity.

Examining Eq. (1), it is seen that the effect of changes in h_0 is dependent on the value (level) of θ and vice versa. Therefore if h_0 and θ are used as control factors, a nonzero interaction effect is anticipated. However, the greater complexity of Eq. (2) is an indication of a poor choice of factors. Notice in particular the strong intermixing of h_1 and h_2 throughout the equation. A very strong interdependence of the factors' effects (i.e., a strong interaction) is likely.

Figure 14 shows effects calculated using these two equations for the two possible sets of control factors. Note in particular the relative magnitude of the effects (Δ) obtained for the factors vs. the interaction. As anticipated, an interaction does occur between h_0 and θ, but it is relatively small. In contrast, the interaction between h_1 and h_2 is very large. By making a poor selection of control factors, a very large interaction has been induced.

Before we get too carried away with this approach, however, a reality check is necessary. This example is based on the application of a mathematical model for the system under study. But, of course, if a completely accurate model was available, experimentation would not be necessary. Large interactions can and do occur. Modeling and physical insight can and should be used to "improve the odds," but assumptions about interactions being zero (or small) are just that—assumptions unless verified experimentally.

In addition to control factor selection, interaction strength can also be affected by the selection of the characteristic response for experimental analysis. This is discussed in Chap. 5.

G. Resolution

Since the occurrence of strong interactions cannot be ruled out, it is important that we set up the experimental array to minimize the potential for

Source	$\Delta = m_{+1} - m_{-1}$
h	+4.9
θ	-11.0
h x θ	-1.1

(a)

Source	$\Delta = m_{+1} - m_{-1}$
h1	+2.9
h2	+3.0
h1 x h2	+8.0

(b)

FIGURE 14 Effects calculated for catapult experiment (a) with h_0 and θ as control factors; (b) with h_1 and h_2 as control factors.

confounding between factors and interactions. In particular, since lower-order interactions (i.e., those involving few factors) are generally a greater risk, experimental design often focuses on avoiding confounding between factors and low-order interactions. The *resolution* of an experimental design is a rating system for the effectiveness of isolating factors from low-order interactions.

The lowest resolution designation is III. In a resolution III design, at least one factor is confounded with a two-factor interaction. In a resolution IV design, no factor is confounded with a two-factor interaction, but at least one is confounded with a three-factor interaction. In a resolution V design, there is no confounding of factors with two- or three-factor interactions, but at least one factor is confounded with a four-factor interaction, etc. At the extreme limit, there is no confounding and we have a full factorial design.

Array

TC	Columns						
	1	2	3	4	5	6	7
1	-1	-1	-1	-1	+1	+1	+1
2	-1	-1	+1	+1	-1	-1	+1
3	-1	+1	-1	+1	-1	+1	-1
4	-1	+1	+1	-1	+1	-1	-1
5	+1	-1	-1	+1	+1	-1	-1
6	+1	-1	+1	-1	-1	+1	-1
7	+1	+1	-1	-1	-1	-1	+1
8	+1	+1	+1	+1	+1	+1	+1

# of factors	Resolution									
3	Full	Factor placement		A	B	C				
4	IV	Factor placement		A	B	C	D			
5-7	III	Factor placement		A	B	C	D	E	F	G

FIGURE 15 Eight TC Factorial Array from Appendix A.

Column	
1	A
2	B
3	C
4	AxBxC
5	BxC
6	AxC
7	AxB

FIGURE 16 Factor and interaction locations for full factorial design based on 8 TC Factorial Array.

As the number of factors in an array is increased, it becomes more and more difficult to avoid confounding and the best resolution that can be achieved decreases. To illustrate, consider the 8 TC Factorial Array from Appendix A, reproduced in Fig. 15. With three factors, it is seen that the design is "full" (i.e., full factorial), so if factor placement is made, as recommended, there will be no confounding. Location of all of the factors and interactions is available from the table in Appendix A and is reproduced in Fig. 16.

With four factors, the best possible resolution is IV *if factor placement is made as recommended.* Corresponding factor and interaction placement is again found from the table in Appendix A and reproduced in Fig. 17. (For simplicity, only two-factor interactions are shown.)

Column	
1	A
2	B
3	C
4	D
5	BxC, AxD
6	AxC BxD
7	AxB, CxD

FIGURE 17 Factor and two-factor interaction locations for four factors based on recommended factor placements for 8 TC Factorial Array, giving resolution IV design.

Column	
1	A, DxE
2	B, CxE
3	C, BxE
4	D, AxE
5	E, BxC, AxD
6	AxC BxD
7	AxB, CxD

FIGURE 18 Factor and two-factor interaction locations for five factors in an 8 TC Factorial Array, producing a resolution III design.

With five or more factors, nothing better than a resolution III can be obtained. To illustrate, consider a design with five factors placed in the first five columns. There is no table in Appendix A for this case, but one can be developed using the interaction table from the appendix. This is given in Fig. 18 and clearly shows the confounding between factors and two-factor interactions. (Again, only two-factor interactions are shown.)

Figures 16 and 17 show the best possible resolution that can be obtained. However, if factors are not placed as recommended, a lower resolution may be produced. This is why particular factor placements are recommended in Appendix A. To illustrate, assume that an 8 TC Factorial Array is used for four factors (A, B, C, and D), but with the factors placed in columns 1, 2, 3, and 5, respectively. The location of each of the interactions is found using the interaction table. Figure 19 shows the

Column	
1	A
2	B, CxD
3	C, BxD
4	AxD
5	D, BxC
6	AxC
7	AxB

FIGURE 19 Factor and two-factor interaction locations for four factors for 8 TC Factorial Array. Recommended factor placements were not used, producing a resolution III design (compare with Fig. 17).

completed pattern. (Again, only two-factor interactions are shown.) Note that there is confounding between factors and two-factor interactions, making this a resolution III design. Compare this confounding pattern with that in the resolution IV design with the same number of factors shown in Fig. 17. Although the resolution III design might "work" in some cases, the resolution IV design has a superior overall confounding pattern.

H. Location of Interaction Terms (Screening Arrays—Appendix C)

The location of interaction terms for the two two-level "screening arrays" in Appendix C (i.e., the 12 TC and 24 TC Screening Arrays) is different from that of the factorial arrays. In the factorial arrays, the interaction term between any two columns is located in another (single) column. In contrast, for the screening arrays, the interaction between any two columns is distributed, "smeared out," over the remaining columns. This leads to resolution III designs since there is always some confounding of factors and two-factor interactions and generally prevents the use of these designs to measure interaction effects. However, it has the advantage of diluting the influence of a potentially strong interaction on the effects calculated for individual factors in the array. Screening arrays are especially useful when it is desired to test large numbers of control factors with relatively small numbers of treatment conditions. They permit sorting of the factor effects by strength, with relatively little risk that a strong interaction will distort the result.

III. DESIGN TRADE-OFFS AND STRATEGIES

A. Four Design Strategies

Strategies for array design are based on establishing a balance among three fundamental considerations: the degree of confounding (resolution), the number of factors that can be tested, and the experiment size (number of treatment conditions). The three-way nature of the required trade-offs is sometimes a source of confusion. It is, for example, possible to greatly increase the number of factors that can be tested for a fixed experimental size. This advantage, however, must be paid for by trade-offs along the third axis, resolution. In many cases, this trade-off is worthwhile, but it should be made intentionally.

The discrete values obtained for resolution provide convenient cutoff points for describing different array design approaches. In particular, it is useful to think in terms of four basic strategies:

1. Full factorial design. This is the most conservative strategy and permits determination of all factor and interaction effects. On the other hand, it is relatively "inefficient" in the sense that the number of factors that can be tested is small relative to the number of treatment conditions.
2. High resolution (i.e., resolution IV and up). With this strategy, factor effects can be determined by assuming that higher-order interactions are zero. If strong two-factor interactions occur, they will not prevent determination of factor effects and, in some circumstances, can also be determined.
3. Specified interaction (a special case of resolution III). This strategy seeks to measure the effects of a limited number of specified interactions along with the factor effects. It permits both interactions and factor effects to be evaluated with a relatively small number of experiments, but depends on the assumption that potentially important interactions can be determined in advance.
4. Low resolution or "saturated" (resolution III). This strategy permits many factors to be studied with relatively small arrays. However, the heavy confounding typically produced means that interaction effects cannot be measured and must normally be assumed zero in order to determine factor effects.

Each of these strategies is described in more detail, with examples of its application, below. First, however, a visual aid, which is sometimes helpful in keeping track of the various array design trade-offs and options, is introduced.

B. The Experimental Triangle

As an aid to visualizing these trade-offs, we have developed the concept of an *experimental triangle* to schematically describe the available design space. This is shown in Fig. 20. Each side on this triangle represents one of three "ideals." The side labeled "low confounding" corresponds to full factorial (no confounding) designs. The "many factors" side represents the ability to test many factors. The "small size" side represents the use of the smallest number of treatment conditions. Motion away from any side

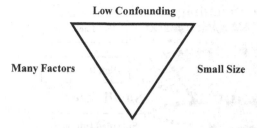

FIGURE 20 The experimental triangle.

corresponds to reduction of that property. For example, moving downward in Fig. 20 corresponds to accepting more confounding in the design.

Any particular experimental design may be represented as a point in this diagram, with placement of the point reflecting the balance struck among the three ideals. The trade-offs involved in changing the design can then be understood in terms of motion of this point. For example, a full factorial design with a small size (e.g., a full factorial design based on the 4 TC Factorial Array) is represented by a point at the upper right corner of the triangle, as shown in Fig. 21. The balance struck in this case is reflected by the fact that this point is far from the "many factors" condition, i.e., only a few factors can be studied. By moving this point (illustrated by the arrows in this figure), the number of factors studied can be increased (i.e., we can move closer to the "many factors" side) but only by increasing the number of treatment conditions, accepting more confounding, or some combination of the two.

The four design strategies, outlined above, are represented by distinct regions on the experimental triangle, as shown in Fig. 22. It can be seen that

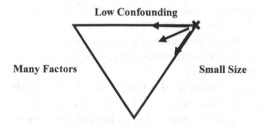

FIGURE 21 Experimental triangle, illustrating the position of a small full factorial design and possible changes to increase the number of factors that can be tested.

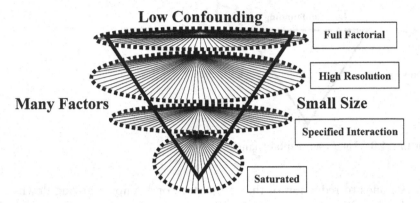

FIGURE 22 Schematic illustration of four experimental design strategies.

vertical motion in the triangle generally corresponds to a shift in the fundamental strategy. In contrast, horizontal motion involves a fairly straightforward trade-off between the number of factors tested and the experimental size.

C. Full Factorial Strategy

This is the most conservative strategy in the sense that it requires no assumptions to resolve confounding and permits determination of all factor and interaction effects. In practice, higher-order interactions are sometimes still assumed to be zero so that their columns may be used for error estimation. The factorial arrays in Appendix A can be used for full factorial designs, but the screening arrays in Appendix C cannot.

The number of treatment conditions required for a full factorial design is given by raising 2 to a power equal to the number of factors to be tested (all factors with two levels). Thus for example, testing four factors requires $2^4 = 16$ treatment conditions. This illustrates the disadvantage of a full factorial design; the number of treatment conditions is large compared to the number of factors that can be tested.

Electron gun optimization: An example of a full factorial study with eight treatment conditions may be found in the work of Felba et al. [3]. In the first portion of this work, the authors studied the effect of three two-level factors (cathode position, diameter of the control electrode orifice, and apex angle) on electron beam emittance. Their experimental design, developed from one of Taguchi's orthogonal arrays, was equivalent to the

8 TC Factorial Array from Appendix A using the recommended factor placement to produce a full factorial. This preliminary experiment showed that the apex angle had such a large effect that it dominated all other terms. The authors therefore performed follow-up experiments with apex angle levels selected closer to the optimum value to further test the effects of other control factors.

Fluid delivery system: A second example of a full factorial design is provided in the fluid delivery study, discussed at length in Chap. 1. Four factors were identified for testing. In addition, it was thought that two-factor interactions (in particular, A × C and B × D) might be important and should be examined. With a relatively small number of factors and a desire to look carefully at interactions, a full factorial design was a good choice. Higher-level interaction terms were also available to provide an error estimate.

The appropriate final design can be assembled from the information in Appendix A. A 16 TC Factorial Array is necessary to get a full factorial. Placing the four factors as recommended, the location of each of the interaction terms is available from the corresponding table in the appendix. Higher-order interactions (i.e., those involving three or all four factors) are assumed zero and the corresponding columns are labeled with "e" to indicate their use as an error estimate. This yields the final design shown in Figs. 3 and 5 of Chap. 1.

To illustrate the trade-offs involved in this design, it is useful to consider some alternatives. With this design, the student group was able to examine all four of the factors they had identified and did not push to include additional factors. However, there was some interest in reducing the size of the experiment and two alternatives were considered. In Fig. 23, the 16 TC full factorial design is represented by the point marked with an X

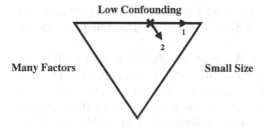

FIGURE 23 Experimental triangle showing two possible options for fluid delivery study as described in the text.

in the experimental triangle. Two possible design changes to reduce the experimental size, i.e., to move closer to the "small size" side of the triangle, are also labeled.

Option 1 involves making a straightforward trade-off between experimental size and the number of factors tested, while retaining the full factorial nature of the design. Specifically, a full factorial design requiring only eight treatment conditions could be obtained by eliminating one of the factors from testing. This option was not pursued because all four factors were thought to have a strong potential for influencing system performance.

Option 2 involves a trade-off between experimental size and resolution, keeping the number of factors unchanged. An 8 TC Factorial array is possible. This reduces the experimental size by a factor of 2. Placing the four factors as recommended gives a resolution IV design. This design is the same as that discussed previously, and the resulting confounding pattern may be seen in Fig. 17. There is confounding among two-factor interactions, making it impossible to estimate them independently. Note in particular that $A \times C$ is confounded with $B \times D$. Assuming that all interactions are zero, this design would permit factor effects to be determined and provide three columns (columns 5, 6, and 7) for an error estimate. However, the group decided that being able to test the two-factor interactions was worth the extra effort involved in the original (16 TC) design so this option was not taken.

D. High-Resolution Design Strategy

The high-resolution design strategy permits some increase in the number of factors that can be tested (compared to the full factorial) without causing confounding of factors with two-factor interactions. If strong two-factor interactions occur, they will not prevent determination of factor effects and, in some circumstances, can also be determined. The factorial arrays in Appendix A can be used for high-resolution designs, but the screening arrays in Appendix C cannot.

Rapid thermal processing: An example illustrating application of a high-resolution design is provided by the work of Amorsolo et al. [4] on the rapid thermal processing of titanium silicide. This example is based primarily on their work. When a titanium film on a silicon wafer is exposed to high temperature, it can react to produce a film of "C54," a titanium silicide phase of interest because of its low electrical resistivity and morphological stability. In rapid thermal processing (RTP), heating of the wafer

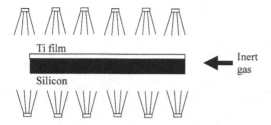

FIGURE 24 Schematic illustration of rapid thermal process (RTP) for producing titanium silicide film. (Adapted from Ref. 4.)

is performed with heat lamps as illustrated in Fig. 24. Other (higher resistivity) titanium silicide phases can be formed during processing. One goal of this study was to identify a set of process conditions (control factors) that produced a fully C54 film for subsequent morphological stability experiments. Five two-level factors were examined (temperature, time, wet etching, sputter etching, and atmosphere). Figure 25 shows the factors and their levels. Relatively strong interaction terms were considered possible (favoring a full factorial design), but it was also necessary to minimize the total number of treatment conditions tested. A high-resolution design provided a reasonable compromise between these two constraints. The experimental design chosen was generated from first principles (i.e., from scratch), but was equivalent to the 16 TC Factorial Array from Appendix A, with the five factors placed as recommended. This design is resolution V. Factors are confounded with four-factor interactions and two-factor interactions are confounded with three-factor inter-

Factor	Level	
	-1	+1
A) Temperature (°C)	650	750
B) Time (seconds)	30	60
C) Wet etch	no	yes
D) Sputter etch	no	yes
E) Atmosphere	argon	nitrogen

FIGURE 25 Control factor levels in rapid thermal processing study. (Adapted from Ref. 4.)

Column	
1	A, BxCxDxE
2	B, AxCxDxE
3	C, AxBxDxE
4	D, AxBxCxE
5	E, AxBxCxD
6	AxE, BxCxD
7	BxE, AxCxD
8	CxD, AxBxE
9	CxE, AxBxD
10	BxD, AxCxE
11	AxD, BxCxE
12	DxE, AxBxC
13	BxC, AxDxE
14	AxC, BxDxE
15	AxB, CxDxE

FIGURE 26 Factor/interaction locations in 16 TC Factorial Array set up for resolution V, equivalent to the experimental design used in the rapid thermal processing study.

actions. Figure 26 shows the complete confounding structure of this design including three- and four-factor interactions.

Figure 27 shows an ANOVA table based on use of the center sheet resistance as the characteristic response. All of the interaction columns (i.e., columns 6–15 in Fig. 26) have been used for the error estimate. For $P = 95\%$, $F_{cr} = F_{0.05, \, 1, \, 10} = 4.96$, which leads to the conclusion that three factors A (temperature), D (sputter etching), and E (atmosphere) are significant. Note that the other two factors have very small effects. The

Source	SS	dof	MS	F
A (temperature)	13.32	1	13.32	**8.9**
B (time)	0.15	1	0.15	0.1
C (wet etching)	0.04	1	0.04	0.03
D (sputter etching)	46.92	1	46.92	**31.3**
E (atmosphere)	21.76	1	21.76	**14.5**
error estimate	15.00	10	1.50	-
Total	97.20	-	-	-

FIGURE 27 Analysis of variance (ANOVA) table for the rapid thermal processing study. (Adapted from Ref. 4.)

contribution of the error term to the total sum of squares, which also includes the contribution of all of the two- and three-factor interaction terms, is about 15%. Thus a high-resolution design produced very definitive results for the factors in the study, permitting the significant factors and their best levels to be clearly identified.

The one major limitation of the analysis presented so far is that the interaction terms were not assessed for significance. In the original study, this drawback was avoided by using an alternative technique, the normal probability plot, to judge significance. This combination of a high-resolution design analyzed with a normal probability plot can be particularly valuable and should be borne in mind when considering the high-resolution strategy. Normal probability plots are described in Chap. 8.

E.   Specified Interaction Strategy

The specified interaction strategy assumes that we can identify, in advance, those interactions with a strong likelihood of being significant. Factor placement is then performed with the goal of keeping these "specified interactions" from confounding with the factors and with each other. By assuming that the effect of all other interactions is zero, the effects of the factors and the specified interactions can be determined. The factorial arrays in Appendix A can be used for specified interaction designs, but the two-level screening arrays in Appendix C cannot. (The three-level, 18 TC, array in Appendix C is an exception and is considered in Chap. 7.)

It is possible to use this strategy with large numbers of interactions specified. This tends to make factor assignment to columns quite complex and Taguchi has developed special sets of tools ("linear graphs") to make the process somewhat less cumbersome. Picking an extensive set of important interactions from an even larger body of candidates is, however, difficult and problematic. More modestly, the designer may be able to identify a small number of interactions of particular concern that he or she wishes to avoid confounding with the factors. In this case, a suitable design can typically be developed with a minimum of trial and error.

To illustrate the process, assume that an 8 TC Factorial Array is to be used to study five factors (A, B, C, D, and E). Furthermore, assume that there are two interactions, (A × B and A × C) that the designer is particularly anxious to avoid confounding with the factors. Using the table of interaction relationships for this 8 TC Factorial Array from Appendix A (also shown in Fig. 10), a suitable design can be constructed without much trouble. We begin by placing the factor involved in the largest number of interactions (in this case A) in one of the columns (say, column 1). We then place factor B in one of the remaining columns (say, column 2). From the interaction table, A × B (1 × 2) is found to be located in column 7.

We now look for possible columns for C. We cannot choose any of the columns already "occupied" (i.e., 1, 2, or 7) and the interaction of the column we choose with A also cannot occupy any of these. If we place C in 3, from the interaction table, it can be determined that A × C is also in an unoccupied column (1 × 3 = 6), so this works. D and E may then be placed in the remaining columns (4 and 5) giving the final placement shown in Fig. 28.

Not all designs come together this easy, and some sets of factor/ interaction combinations are impossible. However, with a reasonably small number of interactions specified and a little patience, a design satisfying the desired specifications can generally be developed.

Development of an experimental design using this strategy often stops once the specified factors and interactions have each been located in a separate column, as shown in Fig. 28. Unfortunately, this sometimes leaves the erroneous impression that the design methodology has *eliminated* all other interactions from the design. In reality, they have simply been *assumed* to be zero. To illustrate this point, in Fig. 29, an additional column has been added, using the interaction table to locate the positions of all remaining two-factor interactions. The effect measured for column 6, for example, is seen to include the contribution of B × D as well as that of A × C. The ability of this design to measure the effect of A × C is therefore dependent on the up-front assumption that the effect of A × C is much larger than that of B × D.

Column	Specified Factor/Interaction
1	A
2	B
3	C
4	D
5	E
6	AxC
7	AxB

FIGURE 28 Location of factors and specified interactions for specified interaction example.

Column	Specified Factor/Interaction	Other 2-factor interactions
1	A	DxE
2	B	CxE
3	C	BxE
4	D	AxE
5	E	AxD, BxC
6	AxC	BxD
7	AxD	CxD

FIGURE 29 Location of factors and all two-factor interactions for specified interaction example. (compare with Fig. 28.)

Successful use of the specified interaction strategy relies on the designer's in-depth knowledge of the system to permit him or her to pick which interactions (from a generally long list) will be large and which will not. It is, in some ways, the least conservative design strategy since it relies heavily on making very specific assumptions about the behavior of the system under study. There are cases where this strategy is useful, but the user should be aware of the assumptions being made in setting up the design.

F. Low-Resolution ("Saturated") Strategy

This strategy makes use of low-resolution (i.e., resolution III) designs to study many factors with a relatively small number of experiments. Commonly, these are saturated (or nearly saturated) designs, which means that factors are placed in all (or nearly all) of the columns. As a result of the very heavy confounding structure, it is necessary to assume that the interaction terms are all zero in order to estimate the factor effects. Columns in which no factors are placed are commonly used to provide an error estimate. This strategy is distinguished from the specified interaction strategy by the fact that it makes no attempt to single out specific interactions based on their assumed importance—all are assumed to be zero.

The low-resolution strategy may be used with any of the arrays in the appendices. However, to reduce the potential for a single strong interaction to distort the effect measured for a factor, use of the screening arrays in Appendix C is strongly recommended. This strategy is often most useful during the initial study of a system, where there may be too many factors to be economically handled by any of the other approaches. A low-resolution experiment is then performed to "screen" the factor list, identifying the most important ones for further testing.

Injection molded bumper: An example of the use of a low-resolution design is found in a study by Chen et al. [5]. The authors reported that during manufacturing of an injection molded bumper, a problem involving the frequent occurrence of surface defects was encountered. It was suggested that reducing the magnitude of temperature gradients and mold-wall shear stress and reducing their variability might eliminate the problem. For preliminary study, a computer simulation was used to examine the effects of 10 control factors, each with two levels. The experimental design, developed from one of Taguchi's orthogonal arrays, was equivalent to the 12 TC Screening Array from Appendix C. Placing 10 factors in this array leaves one column available. Although there is no experimental error in a computer simulation, there is "modeling error" associated with the assumption that the interaction terms are zero and this column provides some estimate of this. Results of the preliminary screening simulation were used to help set up further simulations, with the effectiveness of the optimized process conditions in eliminating the defects ultimately confirmed by hardware testing.

As shown by the above example, when a fully, or nearly fully saturated, design is employed, factor placement in columns is generally arbitrary. This is especially true when one of the screening arrays is used. It may, however, still be useful to leave several columns empty to act as an error estimate. To illustrate, Fig. 30 shows a low-resolution design developed from the 12 TC Screening Array presented in Appendix C. Eight factors have been arbitrarily assigned to columns. The remaining three columns are left "empty" to serve as an error estimate.

G. Randomization of the Run Order

Before concluding this chapter, it is necessary to touch on one additional point that has been subject to dispute between advocates of different experimental design approaches.

In traditional experimental design, the importance of randomizing the run order (order of execution) of treatment conditions is strongly emphasized. Randomization helps prevent systematic changes in uncontrolled variables from biasing the experimental results. Consider, for example, a factor placed in the first column of the 8 TC Factorial Array (Fig. 15). This factor has a level of -1 during treatment conditions 1–4 and

TC	Columns										
	A	B	e	C	e	D	E	F	G	H	e
1	+1	+1	+1	+1	+1	+1	+1	+1	+1	+1	+1
2	-1	+1	-1	+1	+1	+1	-1	-1	-1	+1	-1
3	-1	-1	+1	-1	+1	+1	+1	-1	-1	-1	+1
4	+1	-1	-1	+1	-1	+1	+1	+1	-1	-1	-1
5	-1	+1	-1	-1	+1	-1	+1	+1	+1	-1	-1
6	-1	-1	+1	-1	-1	+1	-1	+1	+1	+1	-1
7	-1	-1	-1	+1	-1	-1	+1	-1	+1	+1	+1
8	+1	-1	-1	-1	+1	-1	-1	+1	-1	+1	+1
9	+1	+1	-1	-1	-1	+1	-1	-1	+1	-1	+1
10	+1	+1	+1	-1	-1	-1	+1	-1	-1	+1	-1
11	-1	+1	+1	+1	-1	-1	-1	+1	-1	-1	+1
12	+1	-1	+1	+1	+1	-1	-1	-1	+1	-1	-1

FIGURE 30 Possible experimental design for testing eight two-level factors using the low-resolution design strategy and the 12 TC Screening Array. Factors have been placed arbitrarily on columns. Three unused columns ("e") will be used for an error estimate.

+1 during 5–8. Now assume that the experiment is run over several days, during which the ambient temperature steadily rises. If the experiments are run in the original order (1–8), the ambient temperature will be lower when A is at level −1 and higher when it is at +1. The effect measured for factor A will be biased by any effect of the ambient on the response.

In addition to changes in ambient conditions, systematic changes may occur during the course of experimentation as a result of changes in personnel, operator fatigue, wear of the apparatus, drift of the zero between calibrations, etc. Randomization is important in preventing bias from such sources from distorting the results.

In spite of the strong case that can be made for randomization, some advocates of the Taguchi approach to experimental design have discouraged the use of randomization. Two arguments are often advanced in support. First, the experimental effort can be reduced by minimizing the number of times that difficult to change factors need to be altered. There are even graphical aides developed to help place factors into array columns based on how difficult they are to change. Second, the potential for damage, wear, or misassembly of the experimental hardware can be reduced by decreasing the number of times that sensitive hardware must be changed during the experiment. It is also pointed out that even traditional designers will sometimes not randomize the run order for particularly difficult to change factors.

In a sense, the difference here is one of the "default condition." Traditional design uses randomization as the default, making exceptions only on a case by case basis. Advocates of the second approach use nonrandomization as their default and may fail to randomize even when factor changes are easily accomplished.

The authors' recommendation is that randomization should be used whenever possible. Exceptions may be made for particularly difficult to change factors, but should be justified on a case by case basis. In such cases, the experiment should be divided into blocks based on the difficult to change factor and run order randomized within each block. An example of this approach is found in the rapid thermal processing study described above [4]. Because the atmosphere was difficult to change, the experiment was divided into two blocks. The eight experiments with an argon atmosphere were run as one block and the eight with a nitrogen atmosphere were run as a second block. The run order of the eight treatment conditions within each block was randomized.

Homework/Discussion Problems

1) An experimental study is to be run to examine the effects of various process variables on the corrosion resistance of powder metallurgy parts. Five process variables (A, B, C, D, and E) are to be used as the control factors, each with two levels. It is believed that interactions between the factors will be small, except for two specific interactions ($C \times D$ and $C \times E$).

    a. Design a suitable experiment using the 8 TC Factorial Array and the specified interaction strategy.
    b. Prepare a table showing which factors and two-factor interactions will be in each column of your proposed design. (Be sure to include *all* two-factor interactions not just $C \times D$ and $D \times E$.)
    c. What is the resolution of this design? How can you tell?
    d. Develop an alternative design based on use of the high-resolution strategy. Discuss what you would gain and what you would have to give up to use this strategy.
    e. Develop an alternative design based on use of the full factorial strategy. Discuss what you would gain and what you would have to give up to use this strategy.
    f. Sketch the experimental triangle and mark the relative positions of the three designs you have developed.

2) An experimental study is to be run to examine the effects of nine control factors (A–I) on the effectiveness of an automotive exhaust system.

It is believed that interactions between the factors will be small, except for the interactions of A with the six following factors (i.e., A × B, A × C, A × D, A × E, A × F, A × G).

    a.   Design a suitable experiment using the 16 TC Factorial Array and the specified interaction strategy.

    b.   Prepare a table showing which factors and two-factor interactions will be in each column of your proposed design. (Be sure to include *all* two-factor interactions and not just the specified interactions.)

    c.   What is the resolution of this design? How can you tell?

    d.   Develop an alternative design based on use of the high-resolution strategy. Discuss what you would gain and what you would have to give up to use this strategy.

    e.   Develop an alternative design based on use of the low-resolution strategy, using a suitably sized screening array (from Appendix C). Discuss what you would gain and what you would have to give up to use this strategy.

    f.   Sketch the experimental triangle and mark the relative positions of the three designs you have developed.

    3)   In the low-resolution strategy, interactions are assumed to be zero. Assume that the low-resolution strategy is used for an experiment, but that a single strong two-factor interaction is actually present.

    a.   Discuss the consequences of this if a factorial array was used for the design.

    b.   Discuss the consequences of this if a screening array was used for the design.

    c.   Explain why use of screening arrays is generally preferred when using the low-resolution strategy.

    4)   An experiment is to be performed to study the performance of a complex system used for reactive ion etching. During preliminary brainstorming sessions, it is realized that the controls on the machine do not directly set the fundamental process parameters inside the machine. Instead, the process parameters are determined by complex combinations of the machine controls.

    a.   If the control knob settings are to be used as the control factor levels in the experiment, what implications would this have for the design? What strategy would you recommend be used and why?

    b. If you are able to select the control factors to be used in the experiment, describe a possible approach for reducing the likely strength of the interactions.

5) Consider the rapid thermal processing experiment described above. Assume that the number of treatment conditions has to be reduced from 16 to 8.

    a. Develop an alternative design that is still based on use of the high-resolution strategy. Discuss what you would gain and what you would have to give up to make this change.

    b. Develop an alternative design that is based on use of a low-resolution strategy. Discuss what you would gain and what you would have to give up to make this change.

    c. Sketch the experimental triangle and mark the relative positions of the original experimental design and of the two alternative designs you have developed.

6) Consider the injection molded bumper study described above.

    a. Develop an alternative design that is based on use of the high-resolution strategy. Discuss what you would gain and what you would have to give up to make this change.

    b. Develop an alternative design that is based on use of the full factorial strategy. Discuss what you would gain and what you would have to give up to make this change.

    c. Sketch the experimental triangle and mark the relative positions of the original experimental design and of the two alternative designs you have developed.

5

Selection of the Characteristic Response

Overview

In this chapter you will learn how to select and refine the response and the characteristic response to achieve your experimental and project goals. This selection is often done reflexively, based on a seemingly "natural" choice. By considering a few basic principles, it is often possible to refine this choice or identify a different, better, alternative.

The first section of this chapter deals with issues related to basic experimental practice; that is, obtaining the best data for the least cost. In the second section the link between the response selection and fundamental product/process improvement is discussed. The third section describes how the choice of a characteristic response can influence optimization and the implementation of the results. Finally, in the fourth section, Taguchi's approach of using "S/N ratios" is described and analyzed.

Response and characteristic response

Before beginning it is important to review briefly the terminology developed in Chapter 1; in particular the relationship between the "response" and the "characteristic response."

To analyze the data from an experimental array, we need to have a number for each treatment condition, reflecting the effectiveness of the control factor settings applied. This is achieved by measuring the response under different noise conditions and then combining these values to produce a characteristic response for the treatment condition. In the fluid delivery experiment described in Chapter 1, for example, the percentage abrasive in the slurry (the response) was measured at two different points in the process (noise conditions), and the characteristic response obtained by averaging these values.

Developing a characteristic response, therefore, involves two related decisions: choosing the response and determining how to combine the values for different noise conditions to obtain the characteristic response. These two decisions are interwoven in the following sections. However, in general, the first two sections focus primarily on the selection of the response, while the last two are related more to how to "combine" the individual responses to obtain a characteristic response.

The "natural choice"

There is often a seemingly natural choice for the response in an experiment. If you wish to improve fuel economy in a car, miles per gallon may appear to be a natural choice. If the paper in a printer is jamming, choosing the jams per million sheets seems an obvious possibility. If a copier is over-heating, the internal temperature can be tracked. Many times the natural choice is truly the best. In other cases, however, there are better alternatives that can be identified if the experiment is examined from different perspectives. The discussions in this chapter are intended to provide a "library" of useful perspectives.

There is considerable overlap in these perspectives. There are also potential contradictions among them. It is therefore certainly not necessary that the response selected for a particular study addresses all, or even most, of them. Use the individual perspectives to generate ideas. Reject those that are irrelevant or impractical. Work with those that you find useful.

I. GOOD EXPERIMENTAL PRACTICE

A. Easily and Accurately Measurable

This is a criterion that sometimes raises a red flag. "Easily measurable" can conjure up images of experimental apparatus cobbled together with rubber bands and bubble gum. On the other hand, in a laboratory setting it is easy to come up with very complex measurement techniques. These add to the cost and time of conducting the experiment. They can also make it difficult to directly track the results later in a product development process, where use of less sophisticated instrumentation may be necessary. It is important to consider ways of reducing effort and cost without sacrificing accuracy.

A simple example of selecting an easier, but still accurate, technique is provided in the rapid thermal processing study [4], briefly described in Chapter 4. In this study, the underlying goal was to fully convert a titanium thin film into a particular titanium silicide phase. X-ray diffraction could be used to directly examine phase development and was, in fact, used in some of the work. However, analysis of all of the samples used in the experiment would have been very expensive in terms of both time and resources. Instead, as the transformation is known to be accompanied by a reduction in the electrical resistance, measurement of the electrical sheet resistance was employed. This provided an accurate, quantitative way to track the success of the conversion process with much less complex instrumentation.

In measuring the response, we should always choose the simplest and most robust measurement techniques available, consistent with accurately collecting the necessary information.

B. Continuous Rather Than Discrete

Many engineering and design problems are characterized by discrete outcomes. A car either starts or does not. A printer feeds smoothly or stops. A copier reaches thermal equilibrium or overheats. These discrete, pass/fail outcomes are obvious choices for the response. They are likely to be the basis on which the end-user (customer) judges a product. Moreover, for statistical analysis, they can be made "continuous" by performing repeated measurements and determining a rate or probability. It is easy to adopt this viewpoint and try to measure improvement experimentally in terms of reducing the failure rate from, say, 1 in 1000 to 1 in 100,000.

Unfortunately, using a discrete value for the response imposes a heavy penalty. That is, when measurement is made with a discrete system, a great deal of information is discarded. Consider using a response such as feed/fail to measure the performance of a paper feeding system. A sheet that just barely makes it out of the feeder will be rated equal with one that moves smoothly. Similarly, a sheet that does not move at all will be rated equal with one that advances nearly enough to feed. (Recall the discussion of how to define quality in Chapter 2.) This is reflected in the need to perform many pass/fail experiments to determine a single meaningful failure rate.

To avoid this problem, we should try focusing on the underlying (continuous) cause of the ultimately discrete behavior. For example, failure to feed in a paper feeder is often tied to the traction supplied to the paper by the rollers. If the traction is too low it may be unable to overcome friction in the system and paper will cease to feed. By measuring the traction of the rollers the paper sheet feed performance can be tracked on a continuous scale, distinguishing subtle changes in performance obscured in a feed/fail response.

C. Tied Closely to Functionality

The chosen response should have the maximum possible sensitivity to changes in the actual, useful functioning of the system, and the minimum possible sensitivity to nonfunctional variations. High sensitivity to function increases the differences available to detect significant effects in the ANOVA analysis. Low sensitivity to nonfunctional variation reduces the background variability, which tends to obscure effects.

The need for a close tie to function is obvious and seldom entirely missed. By thinking explicitly about sensitivity, however, it may be possible to improve the strength of the tie, and to screen out nonfunctional effects. Responses that are linked to appearance or "feel" are particularly prone to problems in this area and should be scrutinized closely.

Consider for example measurement of the quality of a protective coating. Coating quality is often thought of in terms of the roughness and uniformity of the coated layer. These measures of coating quality are also visually and tactilely appealing. For some protective coatings, however, the effectiveness of the coating is strongly compromised by the occurrence of tiny defects ("pinholes") that can penetrate through even apparently "good" coatings. In such cases, coating roughness may not be a very sensitive response for measuring the actual protection provided by the coating. Pinhole count per unit area or depth are possible alternatives.

II. LINKED TO LONG-TERM IMPROVEMENT

The following perspectives are directed toward making the experimental work most effective in terms of long-term product/process improvement.

A. Measurable Early in the Design Process

Experimental work to optimize a product design may not be very useful if the experiments cannot be conducted until most of the major design choices have already been made. It is important to identify responses that can be measured relatively early in the design process. For example, while the ultimate criteria for judging a printer may be the quality of the prints it produces, "print quality" would generally be a poor choice as a response; not measurable until something approaching a complete system is assembled. Even for modifications to an existing system, the fact that so many individual operations (paper registration, ink flow, absorption, etc.) are required for success means that improvements in any one area may be relatively difficult to track.

To develop alternatives, try focusing on the operation of individual components or subsystems. In experimenting with the function of a particular subsystem, avoid responses that rely on other subsystems or the system as a whole for measurement. For example, to improve the performance of a paper feed tray, consider responses (forces, alignment) measured on the paper as it leaves the tray, rather than those which also depend on its subsequent motion in the printer.

B. Tied Closely to Energy or Mass Flow

Central to many products and processes is input of energy and/or mass in some form and its conversion to another (hopefully more useful) form. For example, a recycling operation may input scrap plastic (and electrical energy) to produce packaging materials or "plastic lumber." An internal combustion engine inputs gasoline and air and reacts them to generate energy. Tracing mass and energy flows is a very effective way to learn how, and how well, a product or process functions.

For this same reason, selection of a response that is closely tied to energy and/or mass flow improves the chances of making long-term, fundamental, process improvements. For example, in precision grinding and polishing operations the energy supplied (in the form of applied forces and relative motion between the part and abrasive media) is used to "convert" the workpiece surface into a particulate form that can be carried away by the coolant. In general, smoother surfaces result from a finer scale removal process, but require more energy. Making very smooth surfaces is not difficult, but making them efficiently and consistently is. The energy required per volume removed (or equivalent) thus provides a very fundamental measure of process efficiency and is among those most important to characterize.

Finally, an additional advantage of taking an energy/mass viewpoint is its usefulness in identifying and selecting control factors. In Chapter 4, the importance of this was illustrated using a catapult and comparing the results produced for two different sets of control factors. Tracing energy/mass flow is often the most effective way of identifying potentially important control factors as well as of identifying [through "models" such as Eq. (4.1)] the best (most fundamental) way of developing a set of control factors.

C. Tracks Function Not Dysfunction

A major obstacle to making long-term improvements in a product or process can be the need to continually fix problems on an ongoing basis. Such "fire-fighting" operations divert time and resources from long-term improvements. Moreover, the fixes tend to be costly and can themselves generate additional problems. For example, to mitigate a problem with a copier overheating, we might consider addition of a fan to reduce the temperature build-up around critical components (e.g., the electronics), tracking the improvement using a parameter such as "temperature of the electronics" as the response. However, there will be some added direct cost (fan and installation). Energy efficiency will also be lowered because, although the fan removes heat from the desired area, it consumes energy and will itself generate some heat. Finally, the fan may generate additional problems, such as vibration or noise, which have to be dealt with. As in the child's game of "whack-a-mole," new problems tend to continuously pop-up replacing the previous set.

Choosing a response which tracks function instead of dysfunction is one way of fighting this tendency. There is generally one designed function, but an infinite number of possible dysfunctions. Tracking the function helps maintain a consistent long-term focus. Moreover, energy and/or mass is generally required for dysfunctions (vibration, heat, scrap). Maximizing the correct (functional) use of energy and mass thus also decreases that available for dysfunctions.

D. Dynamic Responses

The concept of a dynamic response is best understood with reference to the P-diagram discussed in Chapter 2 and, for convenience, reproduced in Fig. 1. It should be recalled that ideal functioning of the product or process

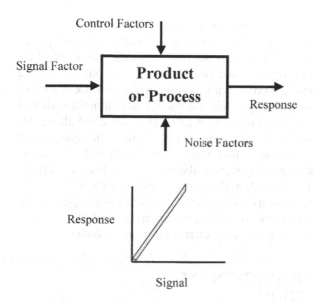

FIGURE 1   The P (product/process) diagram.

involves producing the desired response based on the setting of the signal factor. For example, ideal operation of a furnace system involves producing the desired temperature based on the value set for the power input by setting the furnace controller (the signal).

A dynamic response involves tracking the change in performance as a function of a signal. The alternative, a "static" response, involves determining the response only at a particular value of the signal factor. For example, for the furnace, a static response would be to measure the temperature for a particular setting of the controller. We might set the controller (signal factor) to 500°C and seek to optimize the furnace so that the actual temperature inside the furnace (the response) was as close as possible to 500°C. This is useful when operating near 500°C, but may not produce a design that works well at other settings. In contrast, to measure a dynamic response for the furnace, the temperature at several different signal factor settings (say 200°C, 500°C, and 800°C) is recorded. These data are then used to calculate a dynamic response, such as the slope of the signal vs. the response curve.

Examples of possible dynamic responses include both the slope of a least-squares fit line through the data and the statistical scatter of points about the line. In the case of a thermocouple, for example, the slope of the curve between the actual temperature (signal) and the output voltage is a useful dynamic response. Maximizing this slope will maximize the sensitivity. In other situations the slope itself is of less concern than the scatter of the data about a best-fit line. For a furnace like that described above, for example, the slope of the signal vs. temperature line can generally be adjusted freely over a considerable range ("calibration"). The more difficult problem is reducing the scatter about a best-fit line, so that the internal temperature consistently reflects the controller setting.

Use of a dynamic response will generally require more experimental effort, because it is necessary to make measurements at various signal factor settings, but deliver results of greater long-term usefulness.

III. EASE OF OPTIMIZATION AND IMPLEMENTATION

The following perspectives are directed toward making both optimization based on the experimental results and its subsequent implementation easier.

A. Exploits a Physically Significant Reference or Zero Point

In Chapter 4 we used the example of the motion of a satellite in orbit about the earth, to illustrate the importance of coordinate system selection on the complexity of the resulting mathematical description. In particular, we noted that describing the motion of a satellite in a circular (nongeosynchronous) orbit about the earth is much more difficult using Cartesian $(X-Y)$ coordinates relative to an arbitrary point on the earth's surface than using a polar $(\rho-\theta)$ coordinate system about the earth's center. In Chapter 4 the importance of the coordinate axes (Cartesian vs. polar) was emphasized to illustrate how control factor selection (which sets the "axes" for the experimental analysis) could influence the complexity of the resulting analysis. However, the selection of the reference point is also important. In particular, although the actual motion of the satellite is the same regardless of our choice, we can describe the motion (e.g., as a function of ρ and θ) much more simply using the earth's center as the origin.

One way of identifying such references is to scrutinize points corresponding to extremes (maxima, minima, or natural zero points) in quantities such as the energy, mass, or force. For example, the point of projectile release makes a convenient reference for measuring catapult range [Eq. (4.1)]. In addition to representing a maximum in projectile kinetic energy/ velocity, it is also a natural zero in terms of range. (If the projectile moves backward from this point, there are more fundamental problems than those associated with optimization.)

B. Completeness

Completeness is one of the most difficult criteria to capture. Making a characteristic response "complete" means that its value is correlated on a one-to-one basis with the desired behavior. That is to say, motion of the characteristic response toward its ideal is always linked with actual performance improvement and vice versa. Motion of the characteristic response away from its ideal and performance loss are also similarly linked. A complete characteristic response cannot "trick" us by predicting an illusory improvement in performance or by failing to predict a real loss in performance.

To illustrate the idea of completeness, consider the fluid delivery system discussed in Chapter 1. The characteristic response chosen in this study was the abrasive concentration averaged over the two noise conditions. This is clearly not a complete response. For example, if the average concentration remains the same but there is an increase in the difference in the concentration delivered between the two noise conditions (representing a failure to maintain a consistent concentration) the characteristic response would remain unchanged although the system performance was degraded. Use of average fluid concentration "worked" for the fluid delivery study because preliminary examination of the results identified failure of the system to supply average abrasive concentrations close to the nominal as the main concern. Variability ("dispersion") about the average was judged to be of less immediate importance.

The characteristic response for the fluid delivery study could be made more "complete" by defining it with a function that includes both the average and dispersion. For example, the coefficient of variance ($C_v = \sigma/\mu$, see Chapter 3) might be used. Both increasing the average concentration (μ) and reducing the dispersion (σ) will drive C_v downward. Hence the system could be "optimized" by selecting control factor settings that reduce C_v.

Use of a single function to capture system performance simplifies optimization and implementation, by allowing the experimental team to focus on a single parameter. However, by mixing different physical phenomena (e.g., average and dispersion), it can obscure the underlying physics. Moreover, selection of a particular function carries with it an implicit assumption about relative weightings. For example, use of the coefficient of variance, C_v, weights proportionate changes in the average and dispersion equally. Hence control factor settings which produce large reductions in the average could still be considered "optimum" if they also produce even larger reductions in the dispersion.

In summary, the concept of a "complete" characteristic response represents a useful ideal. However, there are also potential disadvantages if achieving completeness involves combining different physical phenomena into a single function.

C. Optimized by Driving to an Extreme Value

Both optimization, by identifying the best control factor settings, and subsequent implementation, by maintaining the optimized condition during manufacturing and use, are inherently more implementable when the ideal is an extreme value. "Extreme" in this context means either zero (provided negative values are not possible) or infinity.

To illustrate, consider using a group of people to manually transfer jelly beans into a container at a high, but fixed, rate. Clearly, to get close to the desired rate it will be necessary to somehow coordinate their actions. If the rate is too high, performance could be improved by asking some of the group to change their procedure (move more slowly, take smaller handfuls, etc.) so that their contribution is reduced. Conversely, if the overall rate is too slow, we will need to ask for changes in the opposite direction. Under these conditions it is impossible to uniquely identify the "best setting" for any of the individuals involved, as this will depend on the performance of all of the other group members. Moreover, if the performance of any of the individuals involved later changes, it will be necessary to reoptimize. In effect, shooting for an intermediate value as an ideal has "induced" interactions into the optimization process, although there is no actual physical interaction.

Contrast this with the situation that occurs when we have an extreme value as the goal. Say, for example, maximizing the transfer rate (goal of ∞). Optimization is straightforward and the best procedure for each individual contributor is independent of the performance of the others. Moreover, if there is some future change in the performance of one of the group, it does not affect how the others should act. Therefore we do not need to reoptimize.

In some cases selection of a characteristic response with an extreme optimum is obvious. For example, we may wish to maximize the strength of a material or the sensitivity of a sensor. In other cases it is not. As an example, a real manufacturer might be concerned about maintaining a fixed fill rate, so that each container receives the same number (or weight) of jelly beans. It might be easier to figure out how to maximize the rate, but that is not the problem of interest. The question then is whether there is any way of converting a problem that actually involves an intermediate goal into one that can be analyzed in terms of an extreme.

The fluid delivery study provides one example of how this can be done. Ideally, the percentage of abrasive supplied to the polisher would match exactly the nominal concentration, a fixed rather than an extreme value. However, it was soon recognized that failure to supply enough abrasive, due to settling of the abrasive in its bottle, was by far the most serious concern. Any oversupply of abrasive above the nominal concentration was small and to a large extent self-limiting and self-correcting. Thus it was possible to simplify the analysis by concentrating on maximizing the concentration delivered.

The fluid delivery example is a very case-specific solution. However, there is a broad class of problems for which there is a systematic methodology that produces a characteristic response that is optimized at an extreme. These are problems in which reduction in variability is the predominant issue. Two-step optimization, described next, is an approach for analyzing such problems.

D. Two-Step Optimization

Two-step optimization provides a methodology for approaching problems in which variability, rather than average value, is the crucial issue. To illustrate this approach, consider the problem of hitting bull's-eyes in archery. Assume that the initial results are as shown in Fig. 2a. At first

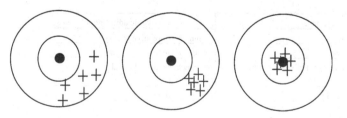

FIGURE 2   Archery example, illustrating the application of two-step optimization.

glance this appears to be a problem involving hitting an intermediate goal (the bull's-eye). With some reflection, however, the problem can be broken down into two components: reducing the spread in the results (i.e., improving consistency) and adjusting the average value so that it is centered on the target. As anyone who has actually attempted this can verify, improving consistency is the hard part. Once you can consistently hit some point on the target, adjusting the impact point to coincide with the bull's-eye is relatively easy. As illustrated in Fig. 2b and c, two-step optimization focuses initially on reducing the variability (without being concerned about "hitting the bull's-eye") and then adjusts the average value onto its ideal.

Identification of a suitable adjustment factor is critical to being able to make this technique work. Adjustment factors may be considered as a special subcategory of control factors. As with all control factors the designer specifies their setting to optimize performance based on the experimental results. However, specification of the adjustment factor is reserved for the second stage of the optimization and is used to place the final average result onto the ideal value, after the other control factors have been set.

An ideal adjustment factor should have a large effect on the average value, but no effect on the variability or the cost. To the extent that the adjustment does affect the variability, the optimum found using the two-step method may be inferior to that possible using a single-step process that considers average and variability together. Therefore it is important to identify an adjustment factor with a minimum effect on variability.

Likely adjustment factors can often be identified by inspection. For the archery illustration, for example, the adjustment factor is the aiming point. For the jelly bean manufacturer, once he/she has established a consistent filling rate, filling time is a probable adjustment factor to put the number/weight of beans per container onto the desired optimum. The table in Fig. 3 provides some additional examples. Adjustment factors can also

response	possible adjustment factor
impact point of arrow	aiming point relative to target
final size of cured resin part	initial (uncured) size of part
stiffness (deflection vs. force) of car accelerator pedal	"spring constant" of pedal
thickness of sputtered film	sputtering time

FIGURE 3   Some possible adjustment factors.

be identified (and likely candidates checked) based on the experimental results. A good adjustment factor should have a large effect on the average, but little or no effect on the variability.

When applicable, two-step optimization is useful in focusing effort on reduction of variability, which is often the most difficult goal to achieve, and in providing a characteristic response that is optimized at an extreme (zero variability). An additional advantage is provided by the identification of adjustment factors during design. If the system is subsequently shifted off of the desired value, for example by changes required in manufacturing, they are available to adjust it back.

Finally, it is important to realize that two-step optimization is most likely to work when the required adjustment during the second step is small to moderate. If the average value produced during the first stage of optimization is far from the ideal, it may be quite difficult to maintain the reduction in variability during the adjustment. Initial work to get the average in the vicinity of the target is therefore often worthwhile even if we ultimately plan to follow a two-step approach. To extend the archery analogy a little further, we are more likely to be able to retain consistent performance when we shift the aiming point, if the shift involves a relatively small adjustment from the edge of the target to the bull's-eye (Fig. 1). If we start from a position far off the target, it is both more difficult to measure the performance and more likely that the final large adjustment in the aiming point will cause other changes (e.g., stance, aerodynamics) that distort the results.

IV. TAGUCHI'S *S/N* RATIOS

A. What is an "*S/N*" Ratio?

One of the most controversial of Taguchi's techniques for experimental design has been the use of so-called "*S/N*" or "signal-to-noise" ratios. Despite the name not all "*S/N*" ratios involve a ratio of signal to noise. They are, perhaps, best thought of as functions designed to capture some of the ideal characteristics described above in a prepackaged form. "Reverse engineering" of some of the most popular of the *S/N* ratios shows that they generally share four characteristics:

1. "Economic justification." *S/N* ratios attempt to capture the true cost of deviating from ideal performance, generally through the application of the quadratic loss function (discussed in

Chapter 2). In theory, a sufficiently broad interpretation of cost would provide an effective way to approach "completeness." In practice, the quadratic loss function is much more narrowly defined. Nevertheless, it does serve to incorporate both average and variability effects into a single value.

2. If the function defined in 1) is not optimized at an extreme, the problem is redefined and the equation modified so that it is. Generally, this takes the form of applying a two-step optimization procedure similar to that described in Section III D.

3. A log transform is performed. Use of the log is generally justified as a measure to reduce the strength of interactions among the control factors. However, the effectiveness of this is dependent on the nature of the original interaction. If the control factors work together in a multiplicative way, taking the log will make them additive and reduce the strength of the interactions. In other cases, however, use of the log may be ineffective or even counterproductive.

4. The function is multiplied by either $+10$ or -10, as necessary to produce a new function that is optimized at $+\infty$. The factor of 10 is purely cosmetic. For consistency, it is desired that S/N ratios always have their optimum value at $+\infty$. Therefore if the function is already optimized at $+\infty$, $+10$ is used. If, however, the optimum is at $-\infty$, then a -10 multiplier is used to change sign. Recall that multiplying a log by a negative number is also equivalent to inverting the argument.

The structure of S/N ratios is illustrated in the following subsections, which describe some of the more commonly seen S/N ratios.

B. Smaller-the-Better S/N

The smaller-the-better S/N is intended for situations where the ideal value of the response is zero, and the response cannot have negative values. Examples where smaller-the-better S/N ratios could be applied include minimizing the number of surface defects on a coating, the energy required to perform a specified task, the waste produced, etc.

1. Justification. Rewriting the standard form for the quadratic loss function [Eq. (2.4)] to eliminate the proportionality constant gives:

$$Q \alpha \left[(\mu - m)^2 + \sigma^2 \right] \tag{1}$$

where Q is the average quality loss (cost), μ is the average response, σ is the standard deviation in the response, and m is the ideal response. For the smaller-the-better case, the ideal response $m \equiv 0$, so the right-hand side of this equation reduces to:

$$\left[\mu^2 + \sigma^2\right]$$

This term provides the core of the S/N ratio.

2. If the function defined in 1) is not optimized at an extreme, the problem is redefined and the equation modified so that it is. In the smaller-the-better case, the function is optimized at an extreme ($m = 0$), so no change is necessary here.

3. A log transform is performed. This gives:

$$\log_{10}\left[\mu^2 + \sigma^2\right]$$

4. The function is multiplied by -10, to produce a new function that is optimized at $+\infty$:

$$(S/N)_i \text{ (smaller-the-better)} = -10 \log_{10}\left[\mu_i^2 + \sigma_i^2\right] \qquad (2)$$

where the subscript i denotes the value for the ith treatment condition. The sum of the square of the average plus the square of the standard deviation (variance) is related to the sum of the squares of the individual responses. So the smaller-the-better S/N ratio is also often written as:

$$(S/N)_i \text{ (smaller-the-better)} = -10 \log_{10}\left(\frac{1}{M}\sum_{j=1}^{M} y_{i,j}^2\right) \qquad (3)$$

where the i subscript designates treatment condition, the j subscript denotes summation over all of the noise conditions, and M is the number of noise conditions.

To illustrate the calculation, assume that the number of visible defects per unit area is used as the response in an experiment on coating deposition. For each treatment condition, the number of defects is measured under three different noise conditions. For treatment condition 1 ($i = 1$) the values recorded for the three different noise conditions are 5,

10, and 15. If the smaller-the-better S/N ratio is used as the characteristic response its value may be calculated using Eq. (3) as:

$$(S/N)_1 = -10 \, \log_{10}\left(\frac{1}{M}\sum_{j=1}^{M} y_{1,j}^2\right)$$

$$= -10 \, \log_{10}\left(\frac{1}{3}\left(5^2 + 10^2 + 15^2\right)\right) = -20.67$$

C. Larger-the-Better S/N

The larger-the-better S/N is intended for situations where the ideal value of the response is infinity, and the response cannot have negative values. Examples where larger-the-better S/N ratios could be applied include the power output from a motor, the speed of a copying process, the strength of a material, etc.

Derivation of the larger-the-better S/N is complicated by the fact that we cannot meaningfully substitute in for the ideal value in the quadratic loss equation (i.e., we cannot use Eq. (1) with $m = \infty$). Therefore the larger-the-better S/N is generally derived by replacing $y_{i,j}$ in the smaller-the-better formula with $1/y_{i,j}$:

$$(S/N)_i(\text{larger-the-better}) = -10 \, \log_{10}\left(\frac{1}{M}\sum_{j=1}^{M}\frac{1}{y_{i,j}^2}\right) \tag{4}$$

D. Signed-Target S/N

The signed-target S/N is intended for situations where the ideal value of the response has an intermediate value, and the average of the response can be adjusted onto the ideal with no effect on the variance. Signed-target S/N ratios can be applied to cases where the response can have negative values. Responses with an arbitrary, adjustable zero (e.g., some voltage, position, and times offsets) are likely candidates for use of the signed-target S/N. The archery problem described in the previous section is an example.

1. Justification. Again begin with the Quadratic Loss Function:

$$Q \propto \left[(\mu - m)^2 + \sigma^2\right]$$

2. Redefining the problem to produce a function optimized at an extreme. In this equation the average response μ can be either

larger or smaller than its ideal value, m, depending on the settings of all of the control factors. As described in Section 3, this makes optimization more difficult as the "best" setting for each control factor is dependent on the settings of all the others. To eliminate this problem two-step optimization is applied. Assume that a suitable adjustment factor can be identified. During the second step of the optimization process, this adjustment factor can be used to move μ onto its target (m), with little or no effect on the variability (σ) or the cost. Hence after adjusting the average onto its ideal value ($\mu = m$) in the second step of the optimization, the quality loss is given by:

$$Q_a \propto \left[(m - m)^2 + \sigma^2 \right] = \left[\sigma^2 \right]$$

where Q_a denotes the quality loss after adjustment. The best setting for any factor is, thus, the one which reduces σ^2, irrespective of the settings of the other factors (assuming there are no interactions involved).

3. A log transform is performed. This gives:

$$\log_{10} \left[\sigma^2 \right]$$

4. The function is multiplied by -10, to produce a new function that is optimized at $+\infty$:

$$(S/N)_i \text{(signed-target)} = -10 \, \log_{10} \left[\sigma_i^2 \right] \tag{5}$$

where the subscript i denotes the value for the ith treatment condition.

E. Nominal-the-Best *S/N*

The nominal-the-best *S/N* is intended for situations where the ideal value of the response has an intermediate value, and an adjustment factor can be used to adjust the average of the response onto the ideal with no effect on the coefficient of variance ($C_v = \sigma/\mu$). The nominal-the-best *S/N* ratio cannot be applied if the response can be negative. Otherwise, the derivation of the nominal-the-best *S/N* parallels that for the signed-target, but assumes that the standard deviation will scale with the average during the adjustment stage. This is often a better assumption for cases that involve absolute changes in distances, dimensions, time intervals, etc. As an illustration, consider the range of a catapult. The shorter the distance the projectile travels, the less scatter anticipated. But ultimately the goal is to minimize the scatter at some fixed range. In the nominal-the-best methodology this is accomplished by first minimizing the σ/μ ratio and then adjusting onto the target.

1. Justification. Again begin with the Quadratic Loss Function:

$$Q \alpha \left[(\mu - m)^2 + \sigma^2 \right]$$

2. Redefining the problem to produce a function optimized at an extreme. In this equation the average response μ can again be either larger or smaller than its ideal value, m, depending on the settings of all of the control factors, complicating optimization. To eliminate this problem two-step optimization is applied, with the assumption that the adjustment factor can be used to move μ onto its target (m), with little or no effect on the coefficient of variance ($C_v = \sigma/\mu$) or the cost. In this case, however, the adjustment does affect the standard deviation. Specifically, the standard deviation after adjustment scales with the ratio m/μ. After adjusting the average onto its ideal value in the second step of the optimization, and accounting for the proportionate change in σ, the quality loss is given by:

$$Q_a \alpha \left[\left(\left(\frac{m}{\mu} \right) \mu - m \right)^2 + \left(\left(\frac{m}{\mu} \right) \sigma \right)^2 \right] = m^2 \left[\frac{\sigma}{\mu} \right]^2$$

where Q_a denotes the quality loss after adjustment. As m^2 is a constant, it will not affect the optimization, so it may be dropped, giving:

$$Q_a \alpha \left[\frac{\sigma}{\mu} \right]^2$$

The best setting for each control factor is, thus, the one which reduces the ratio σ/μ, irrespective of the settings of the other factors (assuming again that there are no interactions involved).

3. A log transform is performed. This gives:

$$\log \left[\frac{\sigma}{\mu} \right]^2$$

4. The function is multiplied by -10, to produce a new function that is optimized at $+\infty$.

$$-10 \log_{10} \left[\frac{\sigma_i}{\mu_i} \right]^2$$

where the subscript i denotes the value for the ith treatment condition. For the nominal-the-best S/N it is common to invert the argument by eliminating the negative sign, creating a true "signal" (μ) to "noise" (σ) ratio for this case:

$$(S/N)_i(\text{nominal-the-best}) = +10 \log_{10} \left[\frac{\mu_i}{\sigma_i} \right]^2 \qquad (6)$$

F. S/N Ratio Application

An example of the application of a Taguchi S/N ratio to an industrial problem may be found in the work of Whitney and Tung [6]. In this study a response for improving the consistency/predictability of the surface left by a grinding process was developed. Responses for multiple points on the ground surface were then incorporated into a suitably chosen Taguchi S/N ratio for analysis.

Response: In the process being studied, a ball end mill was used to establish the basic form/shape of a stamping die. This left a surface consisting of rounded grooves separated by cusps, which were then ground off to smooth the surface. If the grinding process left perfectly flat areas on the cusps, the amount of material removed could accurately be determined by a machine vision system. However, any rounding of the ground "flats" introduced uncertainty. The authors measured the height of the cusp in the middle and on both sides and converted the difference between the values into an area variation (uncertainty). To adjust for geometrical changes caused by the shape of the cusps this was divided by the area already removed by grinding. Thus the chosen response was:

$$y = \frac{\Delta A}{A_r}$$

where ΔA was the variation in the removal area and A_r the area removed by grinding.

Characteristic Response: For each treatment condition, values of y were measured at three separate locations, representing the noise conditions for this study. This is an example of using "repeated measurement" to capture noise effects, as described in Chapter 6. As the response was a quantity with an ideal value of zero, it fit the smaller-the-better S/N case. Thus substituting in for y in the smaller-the-better formula [Eq. (2)], the characteristic response for each treatment condition was obtained as:

$$(S/N)_i(\text{smaller-the-better}) = -10 \log_{10}\left(\frac{1}{3} \sum_{j=1}^{3} \left(\frac{\Delta A_{i,j}}{(A_r)_{i,j}} \right)^2 \right)$$

Control factor settings to optimize the grinding system were then selected based on which level produced the largest S/N ratio for each of the factors judged as significant.

G. Final Comments on Taguchi S/N Ratios

As mentioned at the beginning of this section, Taguchi's S/N ratios have been among the most controversial of his methods. As can be seen from the above descriptions, S/N ratios tend to incorporate a number of features which make optimization easier. They are always optimized at an extreme ($+\infty$), are relatively "complete" (in the sense that they account for the importance of both average and variability in one term), and incorporate a two-step optimization process where applicable. On the other hand, the ratios contain some essentially arbitrary elements and, by mixing different physical phenomena (e.g., average and dispersion), their use can tend to obscure the underlying physics. Prospective users need to weigh the relative advantages and disadvantages of the "S/N ratio" approach in light of their specific project goals and objectives.

Homework/Discussion Problems

1)   One of a company's major product lines is fluid pumps (Fig. P.1). Performance of the current pumps has been adequate, but not outstanding. Compared to competing pumps two areas have been of particular concern:

The pumps are relatively noisy (produce audible noise).
The pumps have relatively high electrical power consumption.

Parameter design experiments are to be performed to improve the performance of the next generation of pumps.

FIGURE P.1 Illustration for problem 1.

It has been proposed to initially concentrate on the audible noise problem during these experiments. Under this proposal, a microphone assembly will be used to measure the audible noise, with the peak value during pump operation used as the response. The engineer making this proposal points out that this response is both continuous and accurately and easily measurable.

    a. Discuss this proposal in terms of the perspectives presented in this chapter.

    b. Propose an alternative choice for the response. Be specific. Discuss/explain your choice in terms of the perspectives presented in this chapter.

  2) Reconsider the selection of the average concentration as the characteristic response for the fluid delivery example from Chapter 1.

    a. Does this selection make use of a "natural zero"? Explain.

    b. In Chapter 1 the decision to focus on the average concentration first and then consider the variability was discussed. This approach seems to run opposite to the logic of the two-step optimization process described above. Discuss.

    c. Consider how the fluid delivery system might be used in actual machine operation. Identify a possible signal factor and use it to propose a dynamic version of the characteristic response.

    d. It is suggested that the final quality and consistency of the polished optics be used as the characteristic response instead. Critique this approach based on the discussion in this chapter.

  3) Analyze the functioning of an automotive exhaust/muffler system from the perspective of energy and mass flow. Define the "function(s)" of the system in terms of mass and energy conversion. Use this analysis to propose a response to improve and track system performance. Justify your proposal.

4)   Assume that you have been assigned to the design team for a new coffee maker. Propose a measurable response to improve and track system performance. Justify your proposal.

5)   Cereal boxes in an assembly line are supposed to be filled with a minimum of 12 oz. of corn flakes. Because of variability in the weight of cornflakes placed in each box, it has proven necessary to set the average weight to 12.2 oz to ensure that the minimum is met.

    a.   Is this problem a likely candidate for the use of two-step optimization? Explain why or why not.

    b.   If two-step optimization is used, propose a possible adjustment factor.

6)   "Noise" is used with multiple meanings in experimental design. Sketch the "P-diagram" and use it to explain the relationship between "noise" as expressed in the noise factors and "noise" as expressed in a S/N ratio.

7)   Reread the problem description for problem 3-4. This problem is to be reanalyzed, using an S/N ratio as the characteristic response. Figure P.2 provides information identical to that given in problem 3-4, except that two additional rows have been added. The first gives the standard deviation of the responses measured for each treatment condition. The second provides space to enter an S/N ratio.

    a.   Select the best S/N ratio for this case from those discussed in this chapter. Explain your choice.

    b.   Calculate the S/N ratio for each of the treatment conditions and enter it in the table above.

    c.   Conduct an ANOM analysis using the S/N ratio as the characteristic response. Present your results in both tabular and graphical form. Determine the "effect" for each factor and identify the "best setting."

    d.   Conduct an ANOVA analysis using the S/N ratio as the characteristic response. Prepare an ANOVA table in the standard form. Judge the significance of the five factors and rank them in order of importance. (Use $P = 90\%$.)

    e.   Would your answer to part d) differ if you had used a different value of P? Discuss.

    f.   Are there any signs of trouble in the ANOVA table prepared for d)? Discuss.

    g.   Predict the S/N ratio that would be obtained using the best settings (i.e., levels) of all of the significant control factors.

TC	A	B	C	D	E	e	e	μ_t (mm)	σ_t (mm)	S/N_i
1	-1	-1	-1	-1	+1	+1	+1	17	1.2	
2	-1	-1	+1	+1	-1	-1	+1	18	1.1	
3	-1	+1	-1	+1	-1	+1	-1	20	0.9	
4	-1	+1	+1	-1	+1	-1	-1	21	1.2	
5	+1	-1	-1	+1	+1	-1	-1	23	1.2	
6	+1	-1	+1	-1	-1	+1	-1	29	0.9	
7	+1	+1	-1	-1	-1	-1	+1	28	0.8	
8	+1	+1	+1	+1	+1	+1	+1	25	1.0	

FIGURE P.2 Data for problem 5.

8)   The output voltage is being used as the response to measure the performance of an electrical circuit with 5 V being the ideal response. It is possible to shift the average voltage merely by adjusting the reference. This adjustment has no cost and no effect on the variability (σ).

Which Taguchi S/N ratio would be most appropriate for this situation?

9)   One simple way of assessing performance is to measure the fraction of parts that in some way do not meet specifications. In general, use of "fraction defective" as a response is discouraged because it is a dysfunction and not continuous. However, there may be cases where its use is unavoidable.

Derive a Taguchi-type S/N ratio for the fraction defective case, using the approach we have outlined in class.

To begin note that the cost per good part produced $\propto (1-p)/p$, where p is the fraction of defective parts produced. (Hint: Use this expression in place of Q.)

6

Inclusion of Real-World Variability

Overview

In this chapter, you will learn how to incorporate the effects of variability into an experimental design using the concept of noise factors. The first section describes what noise factors are and how, in a general sense, they are integrated into an experimental design. The goals and trade-offs involved are also discussed. The second section provides a menu of specific strategies for incorporating noise factors, each with its own set of advantages and disadvantages. Examples drawn from the technical literature are presented to illustrate these strategies.

I. NOISE CONDITIONS

A. Role of Noise Conditions

Before beginning, it is important to briefly review the terminology developed in Chap. 1, in particular the role of noise factors and conditions. Noise factors are parameters too expensive or difficult to control in the field, but which will be carefully controlled during experimentation. It is important that such parameters be included in the experimental design so that recommendations based on the results will be valid under more realistic conditions. For experimental convenience, it is useful to have a designation to refer to specific combinations of noise factor levels. These combinations are designated as "noise conditions" (NC).

Because noise conditions, by definition, are difficult to control in the field, the goal is generally not to explicitly analyze their effects on the response. Instead, the experimental design is arranged so that all treatment conditions "see" the same set of noise conditions. This helps ensure that our selection of "best" control factor settings are "robust" against changes in the noise conditions. Hence for each treatment condition (set of control factor settings), the response is measured under different noise conditions. These responses are then combined to produce a single characteristic response for the treatment condition.

In the fluid supply experiment described in Chap. 1, for example, a single noise factor was used, the amount of fluid used before measurement. Two levels were chosen, representing 10% and 70% fluid use. (Note that with only one noise factor, there is no need to distinguish between a noise condition and noise factor.) For each treatment condition, the response (percentage of abrasive in the slurry) was measured at each of the two noise conditions (i.e., noise factor levels), and the characteristic response obtained by averaging these values.

B. Classification of Noise

Noise factors are often classified as belonging to one of the following categories:

1. Unit to unit. Unit-to-unit noise factors involve the real differences that can occur in nominally identical products because of differences in the components or the assembly. For example, variation in the dimensions of components used and the care with which they are aligned and fit together.
2. External. External noise factors (also commonly referred to as outer or environmental noise) involve differences in the operating environment. Examples include the humidity and temperature in the use environment, the frequency of use, skill of the user, etc.
3. Inner. Inner noise factors involve changes that occur in a product with use or time. Examples include product deterioration with wear, corrosion, and fatigue.

Classification of noise factors within these categories does not effect how they are treated in any of the strategies described below. The purpose of classification is primarily as an aid in identifying and screening candidate noise factors. Considering each category in turn can help prevent overlooking important sources of variation that should be included in the designed experiment.

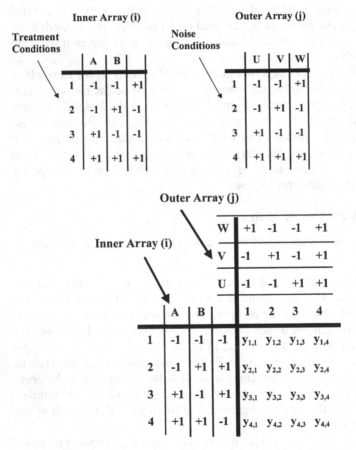

FIGURE 1 Illustration of outer array strategy. Control-factor (inner) and noise-factor (outer) arrays are combined to produce completed design.

C. Goals and Trade-Offs

Before discussing specific strategies for incorporating noise factors into a designed experiment, it is important to consider the goals involved in selecting a specific strategy. The ideal strategy would meet all of the following goals:

- Duplicate "real-world" conditions. Ideally, the noise conditions imposed during testing should simulate the "real-world" conditions as closely as possible.
- Examine performance under extreme conditions. We are often most interested in identifying control factor settings to mitigate against severe degradation of performance, which is most likely to occur under extreme conditions.
- Maximize effects seen in experimentation. Identification of significant factors and best levels is ultimately dependent on "seeing" the real effects against the unavoidable background of experimental and other errors. Selection of noise conditions can aid this by maximizing the experimentally observed differences, and reducing the impact of experimental error.
- Keep the experimental effort as low as possible. As schematically illustrated in Fig. 1, having multiple noise conditions requires collecting multiple data points (responses) for each treatment condition. (Total number of responses collected for the array = number of treatment conditions × number of noise conditions.) If setting each of the noise conditions is laborious, the total experimental effort required can be multiplied to very large values.

Often, these ideal conditions work at cross-purposes. For example, closely duplicating real-world conditions is often facilitated by using a large number of noise conditions, which, unfortunately, tends to increase the experimental effort.

II. A MENU OF STRATEGIES

In this section, we will discuss five different strategies for incorporating noise factors into experimental design, along with some relevant examples.

A. Outer Array

In this strategy, an "outer array" is made from noise factors in a fashion analogous to that used for generating the "inner array" of control factors.

That is, noise factors are specified in terms of discrete levels, an array design is selected, and the noise factors are assigned to specific columns in the array. This outer (noise factor) array is then arranged at a right angle to the inner (control factor) array to set up a TC/NC matrix for responses.

Figure 1 illustrates the concept. In this case, there are two control factors (A, B) and three noise factors (U, V, W) each with two levels. The two control factors are assigned columns in the inner array. This determines the proper setting for each of them for each treatment condition. The three noise factors are similarly assigned columns in the outer array. Thus, for example, noise condition 2 is seen to correspond to factor U at level -1, V at level $+1$, and W at level -1. The outer array is then rotated $90°$ relative to the inner array to set up the matrix specifying the complete factor settings for each response, as shown in the bottom half of the figure. Thus, for example, response $y_{2,3}$, which corresponds to treatment condition 2 ($i = 2$) and noise condition 3 ($j = 3$), would be measured with factor A set to level -1, B to $+1$, U to $+1$, V to -1, and W to -1.

Full outer-array designs are often used to try to capture as balanced as possible an assessment of the effects of real-world variability on the response. As a result, the levels for the noise factors included in outer-array designs are sometimes chosen with the specific goal of producing the same total variance for the noise factor as it has in the real environment. For noise factors with two levels, this is achieved by setting the two levels at the average $\pm$ standard deviation ($\mu \pm \sigma$). For three-level factors, this is achieved with one level at the average minus 1.225 times the standard deviation ($\mu - 1.225\sigma$), one at the average (μ), and one at the average plus 1.225 times standard deviation ($\mu + 1.225\sigma$). To illustrate this idea, assume that the ambient temperature is a noise factor to be included in an outer array. Measurement in the actual environment gives an average temperature of 24°C with a standard deviation of 4°C. To produce the same variance with a two-level noise factor, the levels should be set at 20°C and 28°C (24 $\pm$ 4°C). For a three-level factor, the levels would be set to 19°C, 24°C, and 29°C (24 $-$ 1.225 $\times$ 4, 24, 24 $+$ 1.225 $\times$ 4).

Although the inner and outer-array designs shown in Fig. 1 happened to be based on the same design (4 TC Factorial), this need not be the case. The outer-array design may be selected independently of the inner array. Design of the outer array is less critical than that of the inner (control factor) array, because we are generally interested mainly in the overall effect of the noise factors and not on their individual contributions.

An example illustrating application of an outer-array design is provided in the work of Khami et al. [7]. In this study, the voltage response of an airflow sensor for automotive engine applications was measured as a function of the airflow (signal). Nine control factors with two or three levels were incorporated into the inner array, which had 18 treatment conditions. (This type of 18 TC array is described in Chap. 7.) Unit-to-unit variations, i.e., variations in parts and assembly, were of particular concern, and the authors attempted to capture these effects with 2 two-level noise factors; the orientation of the filter when assembled and the housing (two housings were randomly selected). All combinations of these factors were used, giving four noise conditions and a total of 72 (18 treatment conditions in the inner array × 4 noise conditions in the outer array) different control/noise factor combinations for testing.

Surface-mounting process: A second, and more elaborate, example of the outer-array strategy is provided by Bandurek, Disney, and Bendell. This study looked at methods for improving the robustness of a surface-mounting procedure for electrical components. Five control factors with either two or three levels were used in an inner array based on a modified array with 16 treatment conditions. (Modification of standard arrays to accommodate factors with different levels is discussed in Chap. 7.) Five noise factors with two, three, or four levels each were also included and placed in another specially modified "outer" array with 16 noise conditions. Thus the total number of control/noise combinations tested was 256 (16 treatment conditions in the inner array × 16 noise conditions in the outer array), producing a very large experiment.

Outer-array designs were an important component of Taguchi's original experimental methods, and outer-array designs offer several potential advantages. With care, they can provide a reasonably weighted simulation of actual noise environment. Each treatment condition (i.e., set of control factor settings) is tested against an identical set of noise conditions. And, finally, they are relatively easy to set up and analyze. However, as the above examples make clear, use of an outer array has the potential to produce very large overall experimental designs. In addition, depending on the approach taken in selecting levels and setting up the outer array, they may not fully test the extremes. As a result of these limitations, the use of outer-array designs appears to be declining in favor of other strategies, particularly the "stress test" strategy described next.

B.   Stress Test (Extremes)

In this strategy, all of the noise factors included in the experiment are "combined" to create a small number (typically 2 or 3) of extreme noise conditions. These may represent different types of particularly "bad" (stressful) conditions, but also frequently include a "best" noise condition representing ideal conditions.

Figure 2 illustrates the concept. The "outer array" structure previously described is replaced by a simple listing of noise conditions by number. These can then be separately specified in a listing giving the setting of each noise factor for each condition. Although simple in principle, care does need to be exercised in combining noise factors in this fashion, to ensure that the combinations tested actually represent extremes in terms of the stress they apply to the system. For example, consider combining two noise factors, temperature and humidity, to create two noise conditions.

Noise Factor	Noise Conditions	
	1 (best)	2 (worst)
U	good	bad
V	good	bad
W	good	bad

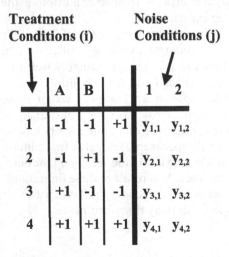

FIGURE 2 Illustration of stress test strategy. Noise factors are combined to give extreme conditions for testing.

Without some understanding of the system, it is impossible to know which of the four possible combinations (cold/dry, cold/wet, hot/dry, hot/wet) should be used to give the best or worse extremes.

In this strategy, noise factor levels are generally chosen to represent relatively extreme conditions. However, care is needed to avoid extremes where the product or process simply cannot work properly, especially given the way that the noise factors are combined. For example, consider choosing levels for five independent (uncorrelated) noise factors such that each represents a 1 in 100 probability of occurrence. In the experiment, the five factors will all have these levels simultaneously; however, the probability of this occurring in application would be only 1 in a 10 billion. Depending on the importance of the noise factors, this may well represent a "perfect storm" that will crush any possible design.

An example illustrating application of stress testing is provided in the work of El-Haiku, Uralic, Cecile, and Drayton [9]. This study sought to improve the press fitting process for a power steering pulley hub and shaft. Eight control factors with two or three levels each were placed in an 18-treatment-condition array. In addition, In addition 13 noise factors, ranging from environmental conditions (temperature, humidity) to part quality, to wear, were identified. For each noise factor, bad and good levels were identified and these were then grouped to derive two noise conditions ("good" vs. "bad"). Testing thus involved measuring the response under 36 (18 treatment conditions × 2 noise conditions) different control/noise factor combinations.

Key advantages of the stress testing strategy include an ability to evaluate the product/process under extreme conditions, and a concomitant tendency to maximize the effects observed. As with the outer-array design, each treatment condition (i.e., set of control factor settings) sees identical noise conditions. The required experimental effort is generally much smaller than that in an outer array, although it is still larger than that required by some of the other strategies. An important concern/limitation on stress testing is its "weighting" of the noise effects, which is generally far from that experienced in the real environment.

C. Surrogate

In this strategy, one or more surrogate factors are used to "capture" the effects of the underlying noise factors. Surrogate factors are used because they are easier to control than more fundamental noise factors and because they permit noise effects to be captured with less experimental effort. They are often position-, age-, or user-based, and can include many "levels." Figure 3 illustrates the experimental layout.

The concept of a surrogate factor is most easily explained using an example. For this illustration, we will return to the rapid thermal processing study first discussed in Chap. 4. The goal of this work was to develop a rapid thermal process (RTP) that could fully convert a titanium thin film on a silicon wafer into the desired "C54" phase. Figure 4a shows the RTP system schematically. The wafer is heated by high-intensity heat lamps, while an inert gas flows over it to prevent contamination. Noise factors identified for this system included variations in the amount of heat output from the lamps, convective loss from the wafer because of the inert gas atmosphere, as well as the radiative heat loss. Lamp output was expected to vary with aging of the lamps as well as run-to-run fluctuations in the power supply and position on the wafer. Similarly, convective loss may vary with run-to-run fluctuations in the amount or direction of gas flow as well as position. Unfortunately, multiple runs with different lamp powers, gas flow rates, etc., although possible, would have been time-consuming and complex. Instead, advantage was taken of the fact that, for a given run, all three of these factors vary across the wafer. This allowed the effects of the noise factors to be captured using the position on the wafer as a surrogate noise factor.

As illustrated in Fig. 4b, for each treatment condition, the response (electrical resistance) was measured at nine points on each wafer, reflecting differences in the amount of energy received from the lamps, and the convective and radiative losses. In executing this design, a large number of

Treatment Conditions (i)			Noise Conditions (surrogate) (j)		
	A	B		1 2 3...	
1	-1	-1	+1	$y_{1,1}$ $y_{1,2}$ $y_{1,3}$ ...	
2	-1	+1	-1	$y_{2,1}$ $y_{2,2}$ $y_{2,3}$ ...	
3	+1	-1	-1	$y_{3,1}$ $y_{3,2}$ $y_{3,3}$ ...	
4	+1	+1	+1	$y_{4,1}$ $y_{4,2}$ $y_{4,3}$ ...	

FIGURE 3   Illustration of array setup for surrogate strategy.

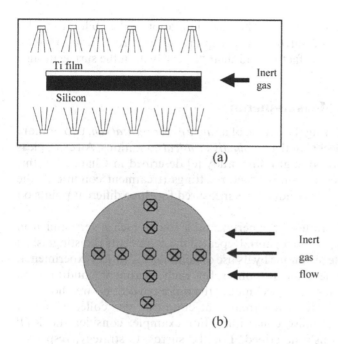

FIGURE 4 Schematic illustration of (a) furnace and (b) testing positions on wafer for rapid thermal processing example. (Adapted from Ref. 4.)

responses were measured (144 = 16 treatment conditions × 9 noise conditions), but it was only necessary to do 16 RTP runs to test the control factor effects. Because the RTP use constituted the main expense in term of both time and money, this technique allowed the experimental effort to be kept down while still incorporating the effects of the noise factors.

The distinction between a surrogate and conventional noise factor can be subtle. In both cases, differences in the response are expected based on changes in the experimental conditions produced by changing the factor level. However, with a surrogate, the differences produced are intended to also capture changes in parameters not being directly varied. In the RTP case, for example, changes in the position on the wafer are used to also include the potential effects of variations in radiative output and gas flow rate, although these are not deliberately varied.

Advantages of the surrogate strategy include reduced experimental effort and an ability to capture noises that are otherwise difficult to control or measure in the laboratory. The main limitation is the ability to find a suitable surrogate or surrogates to capture all important noise factors. Weighting of the noise factors and their "levels" within the surrogate also need to be considered.

D. Repeated Measurements

In this strategy, during the course of a *single experimental run*, two or more responses are collected under *nominally identical* conditions. An example is provided by the robotic grinding study [6] described in Chap. 5. In this study, for each set of control factor settings (treatment condition), the height variation after grinding was measured for three different points on the cusp.

Figure 5 illustrates the experimental layout, which is very similar in appearance to Fig. 4 (surrogate). Repeated measurements is distinguished from the surrogate technique by its use of nominally identical experimental conditions to collect the responses for each treatment condition. In contrast, for the surrogate technique, the noise conditions are chosen so that the responses for each treatment condition are collected under distinctly different noise conditions. For example, consider the RTP experiment previously described. For the surrogate strategy, responses were collected at nine points on the surface. These points were chosen specifically to reflect real differences in the experimental conditions, e.g., difference in the degree of exposure to the flowing inert gas. If, for example, the measurements had instead been made at three arbitrarily selected

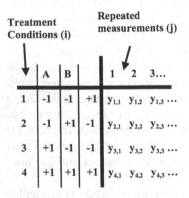

FIGURE 5 Illustration of array setup for repeated-measurements strategy.

points (say three points close to the center of the wafer), it would be an example of the use of repeated measurements.

Use of repeated measurements also needs to be distinguished from replication. Replication is a technique for accurately estimating experimental error (see Chap. 8). In replication, multiple measurements are also made under identical experimental conditions. However, each measurement is made as part of an independent experimental run. In the RTP case, for example, replication would require that several wafers be processed in separate runs for each treatment condition, with measurements on each wafer providing one replicate. This contrast with the repeated measurements, where multiple points are measured for a single experimental run to reduce the experimental effort. Replication is discussed in Chap. 8.

Repeated measurements can be useful in helping reduce the scatter in experimental measurements. For example, roughness measurements on ground surfaces often show considerable random variation from point to point. By measuring each ground surface multiple times, some of the effect of this can be removed from the data. In this case, however, use of repeated measurements may best be thought of as a technique for improving the measurement rather than for incorporating the influence of "real-world variability" (i.e., noise factors).

Use of repeated measurements has the advantage of being easy and inexpensive. However, because random experimental (laboratory) error is, in effect, being used to "capture" noise effects, it is unlikely to provide a good representation of real-world noise conditions. In addition, the effects measured by repeated measurements may be very difficult to observe against the experimental error (which includes the same within-run randomness, as well as the additional effect of run-to-run variations). In some cases, repeated measurements may be the only practical way of obtaining multiple measurements of the response for a single treatment condition. Multiple measurements are required, for example, if a Taguchi style S/N ratio is to be used for optimization. In general, however, this strategy is not recommended as a method for incorporating the effects of noise factors into an experimental design.

E. Single Array (Traditional)

In the single-array strategy, there is no separate set of noise conditions that are used to record multiple responses for each treatment condition. Instead, as schematically shown in Fig. 6, all of the factors are grouped together in a single array, with one response measured for each treatment

Treatment
Conditions (i)   All factors

i	A	U	AxU	Response
1	-1	-1	+1	y_1
2	-1	+1	-1	y_2
3	+1	-1	-1	y_3
4	+1	+1	+1	y_4

FIGURE 6 Illustration of array setup for single-array (traditional) strategy.

condition. Hence the interaction between a control factor (A in Fig. 6) and noise factor (U) appears as a column in the array ($A \times U$). Study of these individual (control/noise) interactions can be used to find control factor settings that reduce the effect of the noise factor on the response. A specific example of this is described below.

The single array is the most traditional approach to experimental design, and can be thought of as representing the opposite end of the spectrum from the outer-array approach. Single arrays are the appropriate design choice when the influence of control (and possibly signal) factors is to be studied in the absence of deliberately introduced noise effects. They can also be used with noise factors, by incorporating the noise factors along with the control factors in the array and examining specific control/noise factor interactions as described above. Finally, single arrays can be useful in working with factors whose classification (noise or control) is not initially clear. If the experimental analysis shows that the factor has a very large effect on the response, it may be necessary to treat it as a control factor and fix its level in the design, even if this is costly or inconvenient. Conversely, if the factor is shown to have little effect, it may be treated as noise and left uncontrolled.

Adhesion tester: A simple example of the single-array strategy is provided in a recent experimental study of an adhesion tester [2]. Because the single-array structure is quite different from the other designs discussed, it is worthwhile working through this example in some detail. The adhesion testing system was built to measure the pull-off force as a function of the maximum applied load. Hence the maximum applied load may be considered as the signal and the pull-off force the response of the system. Experimentation was performed to confirm the dominant influence of load on pull-off force, to check the importance of the ambient temperature as a possible noise factor, and, if necessary, to look for possible ways of mitigating the influence of the temperature.

Experiments were performed with a sapphire indenter contacting a polymer thin film sample. During a typical experiment, the indenter was first loaded with the programmed displacement rate until the load reached the maximum value. This was then held for a preset time and unloaded at a programmed displacement rate to the original position. Five factors were identified for study: temperature, maximum applied load, the rate of displacement during loading and unloading, and the holding time at the maximum load. All of these, except for the temperature, could be quickly and conveniently set for each experiment using the control software. The system permitted control of the temperature in the range from ambient to slightly above, but this required significant waiting periods for system equilibration, making testing slower and more cumbersome. Thus, of the five factors, one (maximum load) was identified as the intended signal factor, and three others (hold time and up and down ramp rates) as control factors. Temperature was flagged for consideration as a possible noise factor, depending on its relative importance.

A full factorial design (32 TC with five factors placed as specified in Appendix A) was used to permit the evaluation of all two-factor and three-factor interaction terms. Columns corresponding to higher-order (four- and five-factor) interactions were used for the error estimate. Levels for the temperature (19°C and 27°C) were selected to "stress" the system, exceeding the extremes normally expected for the ambient temperature in the testing room. Ranges for the other test parameters reflected the desired testing range (load) or reasonable limits on what could be set for use in the final test apparatus and procedure. Information on factor levels is summarized in Fig. 7.

Figure 8 summarizes the ANOVA results from the experiment. For brevity, all two- and three-factor interaction terms that were not judged significant are shown together as "other." The critical F ratio for 99% confidence with 6 degrees of freedom for error and 1 degree of freedom for the factor/interaction is 13.75. (Note that this is misstated in the original text.) Examining the ANOVA table, the error is small and there are no other obvious signs of trouble. All of the factors, except the up ramp rate, and several of the two-factor interactions, are judged significant. The desired signal factor, load, is seen to be the dominant influence accounting

Factor	Level	
	-1	+1
A. Temperature (°C)	19	27
B. Load (g)	1	20
C. Hold (s)	6	600
D. Up ramp (microns/s)	1	3
E. Down ramp (microns/s)	1	3

FIGURE 7 Factor levels for adhesion tester study. (From Ref. 2.)

for almost half the total variance (sum of squares) in the experiment. Moreover, most of the rest of the variance is the result of the control factors, which would be fixed during testing.

The result for the temperature (i.e., noise factor or control factor) is not as clear cut. Temperature produced a sum of squares about 1/6 that of the load and also had a significant interaction with the load (i.e., $A \times B$ in Fig. 8). On the other hand, these were measured over a greater range of temperatures than anticipated during normal operation. To judge whether temperature should be treated as a noise factor or as a control factor, the variability produced by leaving it uncontrolled must be weighed against the inconvenience of controlling it.

Source	SS	DOF	MS	F
A. Temperature	5.553	1	5.553	**267**
B, Load	29.204	1	29.204	**1402**
C. Hold	8.030	1	8.030	**386**
D. Up ramp	0.001	1	0.001	0.04
E. Down ramp	14.729	1	14.729	**708**
AxB	0.875	1	0.875	**42**
AxC	0.359	1	0.359	**17**
BxE	2.371	1	2.371	**114**
CxE	0.797	1	0.797	**38**
"other"	0.916	16	-	-
error estimate	0.124	6	0.021	-
Total	62.960	31	-	-

FIGURE 8 ANOVA table for adhesion tester study. (Adapted from Ref. 2.)

If temperature is considered as a noise factor, its effects can be partially mitigated by making use of its observed interaction with one of the control factors. In particular, use can be made of the observed interaction between temperature and hold time ($A \times C$ in Fig. 8). This is most easily seen by constructing an interaction plot between temperature and hold time. To construct this plot, for each combination of the two factors involved, we calculate an average value using all of the experimental responses with the corresponding factor levels. For example, for the adhesion tester experiment, all of the responses corresponding to 19°C and 6 sec (i.e., for treatment conditions with $A = -1$ and $C = -1$) are averaged. The same is then carried out for the remaining three combinations: 19°C and 600 sec, 27°C and 6 sec, and 27°C and 600 sec. The average response as a function of temperature is then plotted for the two different hold times, as shown in Fig. 9. The existence of a significant interaction between temperature and hold time is reflected in the difference in slope between the two lines in this plot. Notice in Fig. 9 that the line for a 6-sec hold time is significantly flatter than that for a 600-sec hold. Hence, by fixing the hold time at this value, the effects of temperature can be reduced; that is, the test procedure can be made more robust against random variations in the test temperature.

A key advantage of the single-array strategy is its flexibility with respect to factor "type." It can be applied whether or not sorting of factors into "control" and "noise" is possible and appropriate. Thus it is the choice for "traditional" experimentation, where the P-diagram is not applied to differentiate among different factor types. Moreover, even when

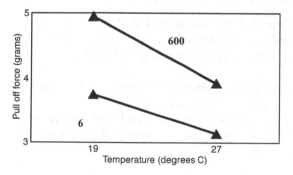

FIGURE 9 Interaction graph (time × temperature) for adhesion tester study. (From Ref. 2.)

control and noise factors are identified, the analysis can be used to produce information about individual control/noise interactions, helping to build a better fundamental understanding of product/process operation. On the other hand, when multiple noise factors are to be analyzed, array design and analysis can be more complicated with this technique. In addition, the single-array technique cannot be used with Taguchi S/N ratios, because their calculation requires that data for multiple noise conditions be recorded for each treatment condition.

Homework/Discussion Problems

1) Describe a recent product or process failure that you feel was caused at least partially by the effect of a "noise factor." What category of noise (i.e., unit to unit, external, or inner) was involved?

2) Assume that you have been assigned to the design team for a new coffee maker. Identify at least six possible noise factors. Include at least one noise factor from each of the three categories described in the text (i.e., unit to unit, external, and inner).

3) The manager of a rock drilling operation is planning experiments to improve the performance of the process for a new project. It is known that rock type has a strong influence on the process performance, and in the new project it will be necessary for the process to perform well in drilling through many different rock types. Samples of a variety of different types of rock are currently available within the laboratory.

    a. Is rock type a "noise factor?" Explain your reasoning.

    b. Explain how the effect of different rock type can be incorporated into the experimental design.

4) To reduce the number of experiments that must be run, an engineer decides to change from an outer-array design to one that includes only two extreme noise conditions. What is being giving up in making this change?

5) Explain the distinction between the surrogate and the repeated measurement strategies. Give an example of each.

6) Compare and contrast the outer-array and single-array strategies in terms of the assumptions made, the experimental effort required, and the complexity of the analysis.

7

Additional Array Designs (Three-Level Factors and Modifications)

Overview

In this chapter, you will learn how to set up and analyze a broader array of designed experiments. In particular, designs for factors with more than two levels as well as those for factors with different levels will be described. The chapter is divided into three sections. The first section presents a set of the most useful designs for three-level factors, paralleling the two-level designs presented in Chap. 4. In the second section, the array design strategies are applied to these arrays with examples drawn from the technical literature. Finally, the third section shows how columns in arrays can be modified to accommodate factors with either more or less levels than those in the standard designs.

I. STANDARD DESIGNS FOR FACTORS WITH THREE LEVELS

A. Basic Trade-Offs for Three-Level Designs

Testing the effects of a factor at three different levels obviously provides more information than that available from two levels. There are, however, fairly steep penalties that must be paid in terms of both experimental size and complexity. The size effect may be seen by considering the number of experiments required to perform a full factorial experiment. This is given by:

number of experiments (full factorial)

$$= \text{(levels per factor)}^{\text{number of factors}} \tag{1}$$

Thus for two factors, a full factorial design requires 9 treatment conditions (3^2) with three-level factors vs. only 4 (2^2) treatment conditions for two-level factors. With three factors, the difference is even larger (27 vs. 8) and continues to grow from there. Very large experiments are relatively rarely used not only because they can be costly, but also because they can be very difficult to manage and execute in a controlled fashion. In this book, we have limited ourselves to designs with a maximum of 32 treatment conditions. This allows us to present a full factorial design for up to five two-level factors (32 TC), but only three three-level factors (27 TC). Many commercial software programs will handle larger designs, and the use of such software, while useful even for smaller designs, grows tremendously in importance as array size and complexity are increased.

A second cost of using three or more factor levels is the potential for higher complexity as a result of the greater information required to characterize interactions. With more levels per factor, there are a greater number of possible combinations of factor levels and this means that more information (i.e., more degrees of freedom) is required. In practice, this means that, for the three-level factorial designs presented here, the interaction between two three-level factors is located in two columns of the array instead of the one required in two-level designs. Moreover, the interaction between three three-level factors, e.g., $A \times B \times C$, is located in four columns. Hence use of both the high resolution and the selected interaction experimental design strategies (described in Chap. 4) is more difficult/limited for three-level arrays.

As a result of these drawbacks, use of two-level arrays is favored in most circumstances and many introductory textbooks present only such arrays. Nevertheless, there are situations where having three levels available for a factor can be important. Often there is an existing "nominal" level for a factor and it is desired to test the effectiveness of values on either side against the nominal. In other situations, we may suspect that the performance is highly nonlinear and multiple levels may be required to assess this. An example is provided by the catapult described in Chap. 4, where the ideal model [Eq. (1) in Chap. 4] suggests a maximum in the range as a function of the launch angle (45° for a launch height of zero). It is not immediately clear how this would be influenced by the noise conditions (differences in projectile shape), but we would probably like to test the response near the predicted maximum, as well as above and below. In situations like this, where this is a possibility of strong curvature, use of a two-level factor could give highly misleading results. Levels chosen on either side of a peak (or valley) in the response might show little or no difference, missing the actual importance. There are also simply cases where we have three or more discrete alternatives that need to be tested and which cannot be captured on a two-level (low/high) scale. For example, there may be three possible types of material, or suppliers, or design variants.

Finally, it is important to note the great importance of one particular three-level based design, the L_{18} popularized by Taguchi, in current industrial practice. This design combines three-level factors with features similar to those described previously for two-level screening designs. Use of this design is described below using the 18 TC (screening) array. The very wide use of this design makes it important that its application be understood.

B. Standard Three-Level Designs

Appendix B provides two factorial arrays using three-level factors, along with various additional information to help in array design. Its use parallels that of Appendix A for two-level designs. A third design based primarily on three-level factors, the 18 TC Screening Array, is provided in Appendix C. Notice that -1, 0, and $+1$ have been used to designate the three levels in the columns. This system is arbitrary and we could instead use 1, 2, 3, or any similar system to make these designations. The order in which these columns are written (i.e., which column is designated #1, etc.) is also somewhat arbitrary, and one may see different ordering in designs generated by different users or different software packages. As with two-level arrays, each three-level factor is placed in a specific three-level column. These positions then determine the location of the interaction between the factors. With three levels, however, the interaction is located in two additional columns rather than one.

To illustrate the use of these arrays, consider first the simplest of the arrays, the 9 TC Factorial Array, and assume that we have two factors to test. Figure 1 shows the first portion of the information presented in the

Array					
TC	Columns				
	1	2	3	4	
1	-1	-1	-1	-1	
2	-1	0	+1	+1	
3	-1	+1	0	0	
4	0	-1	+1	0	
5	0	0	0	-1	
6	0	+1	-1	+1	
7	+1	-1	0	+1	
8	+1	0	-1	0	
9	+1	+1	+1	-1	

# of factors	Resolution					
2	Full	Factor placement	A	B		
4	III	Factor placement	A	B	C	D

FIGURE 1   Nine TC Factorial Array from Appendix B.

appendix. In the upper right-hand corner is the array structure, consisting of four columns. Selecting this array defines the "columns" we will use in conducting and analyzing the experiment. The bottom portion of this table provides recommendations about where to place the factors. With two factors, recommended placement is in columns 1 and 2 of the design. Information below (in the notes and specifications sections) provides additional information about the design. In particular, it can be determined that with this factor placement, the A×B interaction is located in columns 3 and 4 of the array. Figure 2 shows the completed array design for this experiment.

For the second array in Appendix B (i.e., the 27 TC Factorial Array), the structure parallels that for the 9 TC Factorial. However, with the presence of 13 columns, there is additional complexity and the potential for developing high-resolution (and selected interaction) designs. Examples are given below.

The 18 TC Screening Array (Fig. 3) from Appendix C is the last "three-level" design provided. It is functionally similar to Taguchi's $L_{18}(2^1 \times 3^7)$, but with differences in the level nomenclature and ordering of the columns. As with the two-level screening arrays, in this three-level array, the interaction between any two columns is distributed, "smeared out," over the remaining columns. (The important exception to this "smearing out" will be discussed in a moment.) This leads to resolution III designs since there is some confounding of factors and two-factor

TC	Columns			
	A	B	AxB	AxB
1	-1	-1	-1	-1
2	-1	0	+1	+1
3	-1	+1	0	0
4	0	-1	+1	0
5	0	0	0	-1
6	0	+1	-1	+1
7	+1	-1	0	+1
8	+1	0	-1	0
9	+1	+1	+1	-1

FIGURE 2 Nine TC Factorial Array with recommended factor assignments for full factorial design.

interactions and generally prevents the use of these designs to measure specific interaction effects. However, it has the advantage of diluting the influence of a potentially strong interaction on the effects calculated for individual factors in the array. As described above, interaction terms can be particularly limiting when three-level factors are used. Hence this effect is especially useful in a design supporting the use of three-level factors and helps explain the strong popularity of this array in practical application.

Array

TC	Columns							
	1	2	3	4	5	6	7	8
1	-1	-1	-1	-1	-1	-1	-1	-1
2	-1	-1	0	0	+1	0	0	0
3	-1	-1	+1	+1	0	+1	+1	+1
4	-1	0	-1	0	0	0	+1	-1
5	-1	0	0	+1	-1	+1	-1	0
6	-1	0	+1	-1	+1	-1	0	+1
7	-1	+1	-1	+1	+1	0	-1	+1
8	-1	+1	0	-1	0	+1	0	-1
9	-1	+1	+1	0	-1	-1	+1	0
10	+1	-1	-1	+1	0	-1	0	0
11	+1	-1	0	-1	-1	0	+1	+1
12	+1	-1	+1	0	+1	+1	-1	-1
13	+1	0	-1	-1	+1	+1	+1	0
14	+1	0	0	0	0	-1	-1	+1
15	+1	0	+1	+1	-1	0	0	-1
16	+1	+1	-1	0	-1	+1	0	+1
17	+1	+1	0	+1	+1	-1	+1	-1
18	+1	+1	+1	-1	0	0	-1	0

FIGURE 3 Eighteen TC Screening Array from Appendix C.

As shown in Fig. 3, this array has additional unique features not found in any of the other arrays we have examined. Most obviously, it has a mixed structure with a two-level column (column 1) in addition to the other, three-level, columns. As an exception to the "smearing out" described above, the interaction between column 1 (two-level) and column 2 (the adjacent three-level column) can be determined independently of any of the remaining columns. The information (degrees of freedom and sum of squares) for this interaction is "off-the-books" in the sense that it is not represented by a column in this array. However, it can be retrieved. The ability to extract this interaction is also a very useful feature of the design. It permits this array (unlike the other two screening arrays) to be used in a specified interaction design strategy. In addition, it permits us to "combine" the first two columns of this array to create a six-level column for handling a single factor with many levels. The technique of combining columns is discussed in the second section of this chapter.

C. Analysis of Variance for Three-Level Factors

Analysis of variance (ANOVA) for columns with three levels parallels that for two-level columns discussed in Chap. 3.

Degrees of Freedom   For each three-level column, we can calculate three averages, m_{-1}, m_0, and m_{+1}. Since these, in turn, must average to give the overall average, m^*, each three-level column has 2 degrees of freedom associated with it. Factors occupy a single column and so have 2 dof associated with them. For interactions, it is necessary to sum the contributions from each of their columns, so two-factor interactions will have 4 dof (2 dof×2 dof/column) and three-factor interactions will have 8 dof (4 dof×2 dof/column).

Sum of Squares   For three-level columns, the sum of squares (SS) equation shown for two-level columns [Eq. (7) in Chap. 3] can be extended by adding a third term to account for the third level:

$$SS \text{ (three} - \text{level column)} = n_{-1}(m_{-1} - m^*)^2 + n_0(m_0 - m^*)^2$$
$$+ n_{+1}(m_{+1} - m^*)^2 \tag{2}$$

where n_{-1}, n_0, and n_{+1} are the number of characteristic responses with a level of -1, 0, and $+1$, respectively, m_{-1}, m_0, and m_{+1} are the average value of the characteristic response for level -1, 0, and $+1$, respectively, and m^* is the overall average.

To judge significance, the critical F ratio can be obtained from Appendix D and used in the same manner as previously described. However, it is important to realize that because of the different dof associated with factors and interactions, the appropriate value of F_{cr} will depend on which is being judged. For example, consider an analysis where we have three-level factors with 6 dof for error and wish to use a P of 0.95. To judge the factors ($v_1 = 2$), the appropriate value of F_{cr} is $F_{0.05, 2, 6} = 5.14$. To judge the two-factor interactions ($v_1 = 4$), a value of $F_{cr} = F_{0.05, 4, 6} = 4.53$ should be used.

D. Analysis of Variance for the 18 TC Screening Array

For the 18 TC Screening Array, it is also necessary to account for the information (dof and SS) not represented in the form of a column. This is most conveniently done by subtraction. First, contributions from all of the columns are computed and added to give a sum. Then this sum is subtracted from the total to give the "off-the-books" information.

To illustrate, consider the dof calculation. The first column in the array has two levels and thus 1 dof. The remaining seven columns are each three-level, so each accounts for 2 dof. The total in the columns is thus 15 dof $[1 + (7 \times 2)]$. The total dof for entry into the ANOVA table [Eq. (8) in Chap. 3] is $n - 1 = 17$. The difference between these two values gives 2 dof for the noncolumn information ($17 - 15 = 2$ dof).

If there are factors in columns 1 and 2 of the array and the interaction between them is believed to be potentially important, these dof and SS should be entered in the ANOVA table in a row for the interaction. These values are then used to calculate the MS (and thus F) for the interaction in the normal fashion. If, on the other hand, the assumption is made that the interaction is zero, these values can be used in the error estimate.

II. DESIGN STRATEGIES WITH THREE-LEVEL FACTORS

A. Four Design Strategies

Strategies for array design with three-level factors, in parallel with those for two levels, are based on establishing a balance among the resolution of the design, the number of factors that can be tested, and the experiment size. Hence the four basic design strategies, outlined in Chap. 4 and summarized schematically in Fig. 4, are still applicable. There are, however, some differences in emphasis caused largely by the greater "size" and complexity of the interaction terms when three-level factors are involved.

B. Full Factorial Strategy

This remains the most conservative strategy, permitting determination of both factor and interaction effects. With three-level factors, however, relatively few factors can be tested due to the large number of columns required to estimate all interaction effects. The factorial arrays in Appendix B can be used for high-resolution designs, but the screening array in Appendix C cannot.

Factor placement and resulting interaction locations for a full factorial design, generated using the recommended factor placements in a 27 TC Factorial Array from Appendix B, are shown in Fig. 5. A practical example involving the application of this type of design may be found in the work of Reddy et al. [10], who used one of Taguchi's standard arrays to develop the equivalent of a full factorial array. This design was used to study the effect of three three-level factors (concentration, current, and time) on coating thickness produced in a zinc plating process.

Also shown at the far right of Fig. 5 are some simulated data for the sum of squares (SS) and degrees of freedom (dof) for each column in the array. We will use this data to demonstrate the construction and interpretation of an ANOVA table for experiments with three-level factors.

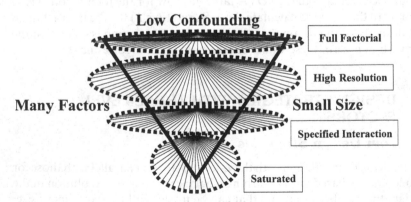

FIGURE 4   Experimental triangle, illustrating four experimental design strategies.

Column	Factor/ Interaction	SS	dof
1	A	200	2
2	B	80	2
3	C	60	2
4	AxBxC	20	2
5	AxBxC	20	2
6	BxC	20	2
7	AxBxC	30	2
8	AxBxC	10	2
9	BxC	20	2
10	AxC	10	2
11	AxC	30	2
12	AxB	100	2
13	AxB	20	2

FIGURE 5 Factor/interaction locations for 27 Factorial Array with recommended factor placement for full factorial design. Right-hand columns (SS and dof) are used in example calculation for ANOVA table (Fig. 6).

Figure 6 shows the ANOVA table, constructed from the data in Fig. 5. For the factors, the SS and dof values are simply entered from the appropriate columns in Fig. 5. Values for each of the two-factor interactions are obtained by summing the values for the two columns in which the interactions are located. Thus for $A \times B$, the SS value is the sum of the values from columns 12 and 13 ($120 = 100 + 20$). Similarly, the dof is the sum of that for the two columns ($4 = 2 + 2$). For the three-factor interaction, values from all four columns should be summed (this interaction is assumed zero and used as the error estimate in Fig. 6). Mean square values

Source	SS	dof	MS	F
A	200	2	100	10
B	80	2	40	4
C	60	2	30	3
AxB	120	4	30	3
AxC	40	4	10	1
BxC	40	4	10	1
error estimate	80	8	10	-
Total	620	26	-	-

FIGURE 6 Analysis of variance table calculation example using data from Fig. 5.

are obtained by dividing the SS value in each row by the corresponding dof entry. Finally, F is calculated by dividing the mean square (MS) for each of the rows by the mean square for error. Values of F_{cr} may be obtained from Appendix D. However, note again that because the factors and interactions are associated with different dof, there is not a single critical F for the entire ANOVA table. For example, with $P = 90\%$, $F_{cr} = F_{0.10, 2, 8} = 3.11$ for judging factor significance. However, with $P = 90\%$, $F_{cr} = F_{0.10, 4, 8} = 2.81$ for judging the significance of two-factor interactions. Thus in the hypothetical case shown in Fig. 6, the interaction $A \times B$ would be judged significant, whereas factor B, with an identical value of F, would not be.

C. High-Resolution Design Strategy

Use of the high-resolution design strategy for three-level arrays is limited by the larger number of columns required for the interactions and the difficulty of fully separating the columns for the factors from those for two-factor interactions. In fact, there is only one high-resolution design listed for the three-level factorial arrays in Appendix B, a resolution IV design that can be achieved for four factors in the 27 TC Factorial Array. The screening arrays in Appendix C cannot be used.

To illustrate this strategy for three-level factors, consider its application to study four factors in a 27 TC array as an alternative to the full factorial design used in the zinc plating study described above. To calculate the factor effects for the high-resolution experiment, it would be necessary to assume that all higher-order (three- and four-factor) interactions are zero. An error estimate could then be obtained by assuming that the two-factor interactions are also zero. Hence the high-resolution design would permit the assessment of one additional factor (four vs. three) at the expense of being unable to evaluate the two-factor interaction terms.

D. Specified Interaction Strategy

As described in Chap. 4, the specified interaction strategy assumes that we can identify, in advance, those interactions with a strong likelihood of being significant. The 27 TC Factorial Array in Appendix B can be used for specified interaction designs. However, because interactions between two three-level factors occupy two columns, use of this technique is more limited than for two-level arrays. In addition, as discussed above, one specific interaction, between the first and second columns, can be estimated for the 18 TC Screening Array from Appendix C.

Recall that in the specified interaction strategy, factor placement is performed with the goal of keeping these "specified interactions" from confounding with the factors and with each other. By assuming that the effect of all other interactions is zero, the effects of the factors and the specified interactions can be determined.

Machine vision An example of a study using the specified interaction strategy is found in the work of Shiau and Jiang [11]. The effect of seven three-level control factors (lens type, background color, on-target distance, to-target distance, filter, lighting source, and angle) on the performance of a 3-D coordinate measuring system was examined (factors A–G). In addition, three two-level factors (two factor combinations of on-target distance, to-target distance, and angle) were specifically identified and the design was conducted to avoid confounding them with either the factors or each other (specified interactions C × D, C × G, and D × G).

In the original work, the experimental design was developed from one of Taguchi's orthogonal arrays. However, an equivalent design can also be generated using the 27 TC Factorial Array from Appendix C. This exercise provides an opportunity to demonstrate the general approach. We begin by placing one of the three factors involved in the interactions (C) in one of the columns (say column 1). We then place Factor D in one of the remaining columns (say column 2). From the interaction table, C × D (1 × 2) is found to be located in columns 12 and 13. Next we need to place the last of the factors involved in interactions (G) in one of the remaining columns (3–11), such that its interactions with C and D are also located in one of the so far empty columns. For example, if we place G into column 3, C×G (1 × 3) will be in columns 10 and 11 and D × G (2 × 3) in columns 6 and 9. The remaining four factors may be placed in any of the remaining columns.

Figure 7 shows the completed design. Each of the factors and the three specified interactions occupies a separate column, and so all may be estimated independently of each other. It is, of course, still necessary to assume that all of the other interactions are zero to do this analysis.

E. Low Resolution (Saturated)

As seen above, for reasonably sized experimental designs, the number of three-level factors that can be studied with full factorial or high-resolution designs is very limited. This makes the low-resolution strategy appear relatively more attractive. Moreover, because interaction terms are assumed zero, these designs are extremely easy to construct even for

Column	
1	C
2	D
3	G
4	A
5	B
6	DxG
7	E
8	F
9	DxG
10	CxG
11	CxG
12	CxD
13	CxD

FIGURE 7 Factor/interaction locations in 27 TC Factorial Array for specified interaction example. Only "specified" two-factor interactions are shown.

three-level factors. Finally, the availability of a conveniently sized screening array (the L_{18} design popularized by Taguchi) provides a way of dealing with some of the effects of the heavy confounding. While this strategy may be used with factorial designs, use of the screening array is strongly recommended.

An example of the application of the low-resolution strategy may be found in the work of Whitney and Tung [6]. These authors studied the effects of five three-level factors (normal force, feed speed, grit size, disc backing, and angle) on grinder performance. The experimental design was constructed from Taguchi's standard L_{18} orthogonal array by simply placing factors in five of the seven three-level columns. This left two three-level and one two-level columns empty. Hence this design has 7 degrees of freedom available to make an error estimate (1 for the two-level column, 2 each for the three-level columns, and 2 for the interaction between columns one and two).

The work of Sano et al. [12] provides a second example of the use of this array. In this case, the effects of one two-level and seven three-level factors (inlet tank length, tube thickness, inlet tank shape, inlet tank flow direction, tube length, tube end shape, inner fin length, and inlet tube inner diameter) on the airflow through an intercooler for use with an automotive turbocharger system were studied. No ANOVA was reported in Ref. 12. However, note that for this particular array, even when factors are placed in all of the columns, the information associated with the interaction between the first two columns still provides a potential error estimate.

III. MODIFICATION OF COLUMNS

A. Motivation and Limitations

The experimental design presented so far, with a single exception, all involved columns (and thus factors) with identical numbers of levels. In some situations, however, we may have one or more factors with a different number of levels. For example, what if the students in the fluid delivery study had wished to test four different stirrer designs instead of two? Modification of arrays does have some important limitations. In particular, it can lead to more complex designs and confounding. Therefore its application tends to be easiest when the "saturated" array design strategy is being used.

This section provides some relatively straightforward techniques for modifying the standard arrays to accommodate factors with a "non-matching" number of levels.

B. Combining Columns Technique

In this technique, two columns from the original array along with the column(s) where their interaction is located are combined to form a single new column with an increased number of levels.

As an example, consider the 8 TC Factorial Array (Appendix A) shown, along with its interaction table, in Fig. 8a. A four-level column may be produced in this array by combining columns. To begin, we select two columns, say 1 and 2. From the interaction table, the interaction between these two columns is located in column 7. Hence columns 1, 2, and 7 are combined to produce a new four-level column, 1*, as shown in Fig. 8b.

Levels in the new column are based on unique combinations of levels from the columns being combined. Thus for example, level 1 in column 1* corresponds to treatment conditions where there was a -1 in column 1, a -1 in column 2, and a $+1$ in column 7. Similarly, level 3 in column 1* corresponds to a $+1$ in column 1, -1 in column 2, and -1 in column 7. The dof for the new column is given by the number of levels in the column minus one. This should also be equal to the total number of dof associated with the columns being combined. For example, in Fig. 8b, the new, 1* column has 3 dof $(4-1)$ which is equal to the sum of those from the old columns 1, 2, and 7 $(1+1+1=3)$.

Interactions of other columns with the new column may be located by finding the location of the interactions with each of the original columns. For example, in Fig. 8b, the interaction between a factor in

TC	Columns						
	1	2	3	4	5	6	7
1	-1	-1	-1	-1	+1	+1	+1
2	-1	-1	+1	+1	-1	-1	+1
3	-1	+1	-1	+1	-1	+1	-1
4	-1	+1	+1	-1	+1	-1	-1
5	+1	-1	-1	+1	+1	-1	-1
6	+1	-1	+1	-1	-1	+1	-1
7	+1	+1	-1	-1	-1	-1	+1
8	+1	+1	+1	+1	+1	+1	+1

(a)

TC	Columns				
	1*	3	4	5	6
1	1	-1	-1	+1	+1
2	1	+1	+1	-1	-1
3	2	-1	+1	-1	+1
4	2	+1	-1	+1	-1
5	3	-1	+1	+1	-1
6	3	+1	-1	-1	+1
7	4	-1	-1	-1	-1
8	4	+1	+1	+1	+1

(b)

FIGURE 8 Illustration of combining columns technique. (a) Original array. (b) New array with original columns 1, 2, and 7 combined to give new column 1*.

column 1* (A) and one in column 4 (B) is located in columns 3, 5, and 6. This can be determined by finding the interaction between column 4 and the three columns that were combined to give 1*. That is, A×B is located in 1×4 (= column 5), 2×4 (= column 6), and 7×4 (= column 3).

For the 18 TC Screening Array, the first two columns may be combined, along with their interaction (not represented by a column) to give a six-level column. In this case, the new column has 5 dof (6−1).

C. Virtual (Dummy) Levels Technique

The virtual or dummy-level technique is used to reduce the number of levels in a column, e.g., to include a two-level column/factor into an array with only three-level columns. To do this, one of the levels in the original column is replaced with a "virtual level" equal to one of the other levels. For example, consider the first column in the 9 TC Factorial Array shown in Fig. 9a. To make this a two-level column, we simply replace all occurrences of the "0" level with the "+1" level, as shown in Fig. 9b. Note that it is important that the same substitution be made everywhere in the column. For example, if "0" is replaced by "+1" somewhere in the column, it is important that all "0"s in the column be replaced by "+1."

When the virtual level technique is used, the symmetry of the array is distorted somewhat. For example, all combinations of levels for two

TC	Columns			
	1	2	3	4
1	-1	-1	-1	-1
2	-1	0	+1	+1
3	-1	+1	0	0
4	0	-1	+1	0
5	0	0	0	-1
6	0	+1	1	+1
7	+1	-1	0	+1
8	+1	0	-1	0
9	+1	+1	+1	-1

(a)

TC	Columns			
	1	2	3	4
1	-1	-1	-1	-1
2	-1	0	+1	+1
3	-1	+1	0	0
4	+1	-1	+1	0
5	+1	0	0	-1
6	+1	+1	-1	+1
7	+1	-1	0	+1
8	+1	0	-1	0
9	+1	+1	+1	-1

(b)

FIGURE 9 Illustration of virtual level technique. (a) Original array. (b) New array with "0" level in column 1 replaced with "+1" level.

columns still occur, but they no longer occur an equal number of times. If factor A is placed in column 1 and B in column 2 in Fig. 9b, at least one treatment combination is run with each of the six possible combinations of A and B. However, combinations involving A at $+1$ occur twice as frequently as those with A at -1.

Reducing the number of levels for a column also reduces the dof for the column. The "missing" dof and an associated SS term are no longer represented in column form. They can be obtained by subtracting the sum of the terms from all columns from the total and are often useful as an error estimate.

Interaction terms involving columns with virtual levels also take fewer dof and therefore do not fully occupy all of the column(s) required before inserting the virtual level. Unfortunately, it can be difficult to simply separate the interaction and error terms using the column-based system for calculations adopted here. A good commercial analysis program is generally able to handle this. These degrees of freedom are often useful for making an error estimate.

D. Use of Both Techniques

The combining columns and virtual level techniques may be used together. For example, to create a three-level column in a two-level array, three columns would first be combined to produce a four-level column. A virtual level would then be applied to the four-level column to convert it to the desired three-level column.

Homework/Discussion Problems

1) Develop a table showing the location of the factors and all of the two-factor interactions for the machine vision example discussed in this chapter.

2) An experiment is to be designed to study the effects of five three-level control factors (A–E). It is believed that interactions between the factors will be small, except for the four two-factor interactions involving factor B (i.e., A×B, B×C, B×D, and B×E)

a. Design a suitable experiment using the 27 TC Factorial Array and the specified interaction strategy.
b. Prepare a table showing which factors and two-factor interactions will be in each column of your proposed design. (Be sure to include *all* two-factor interactions.)

    c. What is the resolution of this design? How can you tell?

    d. Develop an alternative design based on use of the high-resolution strategy. Discuss what you gained and what you had to give up using this strategy.

    e. Develop an alternative design based on use of the 18 TC Screening Array. What strategy must be used in this case? Discuss what you gained and what you had to give up using this array and strategy.

    f. Sketch the experimental triangle and mark the relative positions of the three designs you have developed.

3) Consider the machine vision experiment described above.

    a. Develop an alternative design that is based on use of the high-resolution strategy. Discuss what you would gain and what you would have to give up to make this change.

    b. Develop an alternative design that is based on use of a low-resolution strategy. Discuss what you would gain and what you would have to give up to make this change.

    c. Sketch the experimental triangle and mark the relative positions of the original experimental design and of the two alternative designs you have developed.

4)   An experiment is conducted to evaluate the effect of three three-level factors (A–C) on the number of defects occurring in batches of extruded plastic parts. These data are collected under three noise conditions.

Figure P.1 summarizes the experiment, including the raw data, in the form of the response measured for each treatment and noise condition combination. All interactions have been assumed to be zero. The column containing only interactions is therefore to be used for the error estimate. It is marked with an "e" in the table.

A smaller-the-better S/N ratio is to be used as the characteristic response in analyzing the data.

    a. Calculate the characteristic response (S/N ratio) for each treatment condition and enter them into the table.

    b. Conduct an ANalysis Of Means (ANOM) analysis. Present your results in both tabular and graphical form. Identify the "best setting" for each factor.

    c. Conduct an ANOVA and prepare an ANOVA table in the standard form. Judge the significance of the three factors and rank them in order of importance. (Use $P = 90\%$).

TC	A	B	C	e	Noise Conditions			S/N$_i$
					1	2	3	
1	-1	-1	-1	-1	35	24	18	
2	-1	0	+1	+1	12	15	20	
3	-1	+1	0	0	23	21	22	
4	0	-1	+1	0	27	20	25	
5	0	0	0	-1	35	20	26	
6	0	+1	-1	+1	50	60	40	
7	+1	-1	0	+1	50	60	45	
8	+1	0	-1	0	15	20	25	
9	+1	+1	+1	-1	45	65	40	

FIGURE P.1 Experimental layout and data for Problem 4.

    d. Are there any signs of trouble in the ANOVA table prepared for part (d)? If so, discuss likely causes.

    5) Describe two features of the 18 TC Screening Array that are not present in the three-level factorial arrays.

    6) Starting with the 16 TC Factorial Array, develop an experimental design that can accommodate three three-level and three two-level factors. You will need to use both the combined columns and virtual level techniques and assume that all interactions are zero.

8

Additional Analysis Techniques

Overview

This chapter provides a library of additional analytical techniques for judging the significance of factors and interactions in a designed experiment. Section I reviews and summarizes the technique you have seen so far in this book, i.e., the utilization of "unused" array columns to estimate error for ANOVA analysis. In Sec. II, the widely seen approach of "pooling of smallest effects" is described. Section III describes the traditional approach of replication. Finally, Sec. IV describes the use of normal probability plots, as well as providing a closing discussion and example of some of the other important concepts seen throughout the book.

Choosing an analysis technique

An important theme throughout this book has been the importance of making informed choices in experimental design. At a very basic level, this often involves striking an appropriate balance between experimental effort and the quality and quantity of information obtained. This balance, in turn, is dependent on the specific circumstances such as the project goals, existing knowledge of the system under study, the available resources and time, and the difficulty of the experiments. In selecting an analysis technique, the primary consideration is often the quality of the error estimate. The first three of the techniques described below represent different points along the spectrum of error estimate quality vs. effort. Pooling of unused columns strikes an intermediate balance. Pooling of small effects minimizes effort at the expense of quality of the estimate. Replication provides the best error estimate, but requires the most effort. The final technique (normal probability plots) attempts to overcome the tradeoff by judging significance based on the pattern of all the effects in a data set.

I. POOLING OF UNUSED ("EMPTY") COLUMNS

In previous chapters, we have generally made the error estimate by adding together the dof and SS associated with "unused" columns in the array design. These have typically been columns assigned to higher-order (three- or more-factor) interactions, although two-factor interactions columns have been used when these were not available. This technique often represents a useful compromise between the quality of the error estimate and the experimental effort required to obtain it. Like many compromises, however, it is subject to criticism from both sides.

It is important to realize that what we are calling "unused" columns in the array are not truly empty, in the sense that they represent potentially real effects. These columns generally contain one or more interaction effects that we are assuming to be zero. This assumption may be thought of as a model for the system behavior. Hence to the extent that this model is not accurate, the effects we measure for the column will include a contribution from the "modeling error" as well as the experimental error. This is one of the reasons that, as originally noted in Chap. 3, use of the F ratio test is only an approximation in these applications. Nevertheless, it is still valuable in providing a measure of consistency (and discipline) to judging significance.

Viewed from the opposite perspective, this technique can be criticized as being relatively expensive in terms of its impact on our ability to measure factor and/or interaction effects. In the RTP case study (high-resolution strategy) described in Chap. 4, for example, all of the interaction terms were used to obtain the error estimate shown in the ANOVA table in Fig. 24. For designs employing the low-resolution strategy, each column devoted to the error estimate represents a column that could have been used to test an additional factor.

II. POOLING OF SMALLEST EFFECTS

This technique attempts to avoid the need to set aside columns in the experimental array for error by using the results of the experiment itself to assign columns to the error estimate. If the mean square for a column is small, it may seem reasonable that "there must not be a real (factor or interaction) effect present and, therefore, the column may be used to estimate the experimental error." There is a serious flaw in this reasoning and this has caused considerable confusion and argument. As will be seen in a moment, by being a little more careful in stating the assumptions and goals of the analysis, this confusion can be avoided and the technique usefully employed in some situations. First, however, here is an illustration of the potential flaw.

Figure 1 shows an experimental design in which an 8 TC Factorial Array has been fully saturated by placing a factor in each column. Also shown is a characteristic response (α_ι) for each treatment condition. Figure 2 shows the initial results of an ANOVA analysis of these data. Without an error estimate, we have no way to calculate an F value or judge significance. Looking at the mean squares for the various columns, let us pool the three smallest factor terms (C, F, and G) to obtain an error estimate. This result is shown in Fig. 3. Seeing this table, you might suspect that all four of the remaining factors were significant. Using $P = 90\%$ gives $F_{0.10, \, 1, \, 3} = 5.54$, so factor A would not make the required cut, but B, D, and E would all be judged significant. However, as you

TC	Column							α_ι
	A	B	C	D	E	F	G	
1	-1	-1	-1	-1	+1	+1	+1	47.18
2	-1	-1	+1	+1	-1	-1	+1	55.41
3	-1	+1	-1	+1	-1	+1	-1	29.37
4	-1	+1	+1	-1	+1	-1	-1	33.53
5	+1	-1	-1	+1	+1	-1	-1	25.82
6	+1	-1	+1	-1	-1	+1	-1	56.85
7	+1	+1	-1	-1	-1	-1	+1	34.38
8	+1	+1	+1	+1	+1	+1	+1	16.45

FIGURE 1 Saturated 8 TC Factorial Array with data for characteristic response.

Source	SS	dof	MS	F
A	127.92	1	127.92	
B	639.57	1	639.57	
C	81.22	1	81.22	
D	251.89	1	251.89	
E	351.52	1	351.52	
F	0.06	1	0.06	
G	7.70	1	7.70	
error est	0	0	-	-
Total	1459.88	7	-	-

FIGURE 2 ANOVA table for data from Fig. 1. *F* cannot be calculated without an error estimate.

might suspect from the previous paragraph, there is something "fishy" about the results shown in Fig. 1. Specifically the characteristic responses were actually randomly generated. In spite of the calculation of an error estimate, and the "large" values of *F* produced, none of the factor effects are real. The fundamental lesson is, not to overstate the obvious, if you "pool the smallest effects" to estimate error, the remaining effects will, by

Source	SS	dof	MS	F
A	127.92	1	127.92	4.3
B	639.57	1	639.57	**21.6**
C*	(81.22*)	(1*)	-	-
D	251.89	1	251.89	**8.5**
E	351.52	1	351.52	**11.9**
F*	(0.06*)	(1*)	-	-
G*	(7.70*)	(1*)	-	-
Total	1459.88	7	-	-
error estimate*	88.98*	3*	29.66*	

FIGURE 3 ANOVA table for data from Fig. 1. The three smallest effects have been "pooled" to provide an error estimate (see discussion in text).

definition, be larger than the "error." Depending on the specific values (and how we select terms to pool), very large F values can be generated. But even very large F values obtained in this way are not useful in judging statistical significance.

Of course, the situation setup in the above example is not entirely realistic. In most practical situations, the factors are not simply anonymous letter codes (A, B, C, etc.). Often the strong effect (relative to experimental error) of some or all of the control factors on the measured response can reasonably be assumed in advance. In such cases, the experimental goal is not the determination of whether or not certain "effects" are statistically significant. Instead, it may be to determine the relative importance of the effects. Small effects are then "pooled" together as a group for comparison with the dominant effects. Recognizing this up-front assumption (that the effects observed are real), this technique can be useful.

The term "practical significance" is sometimes used in an attempt to distinguish this type of comparison from the judgment of "statistical significance." However, the similarity of the two phrases can still lead to confusion. Instead, it is probably better to speak in terms of the relative importance or relative contribution when pooling small effects. Similarly, although a value of F is often presented with this technique, this can also lead to confusion. Presentation of results in terms of the percentage contribution to the total SS can help avoid this.

Injection Molded Bumper: An example of the application of this technique may be found in Chen et al. [5]. This study was briefly described in Chap. 4 as an example of the application of a low-resolution design strategy. Recapping, a screening array was used to study the influence of 10 control factors on temperature gradients and mold-wall shear stress during injection molding of an automobile bumper derived using a computer simulation. In this case, there is no "experimental error" as such, but there is "modeling error" associated with the assumption that the interaction terms are zero. The one column with no factor placed in it provides some measure of this. By pooling the smallest effects, an appreciation of the importance of the dominant effects relative to the influence of the small terms (which we may ultimately choose to neglect) is obtained. For example, examining the data in Ref. 5, it is found that the seven smallest effects (six factors and the empty column) account for only about 6.5% of the total variance in the system, compared with about 50% and 33% for the two strongest factors. Assuming that all of the effects are "real" (in this case, this means assuming that they are not the result of "modeling error" caused by neglecting the interaction terms), the dominant role of the strongest few factors is apparent.

III. REPLICATION

Replication delivers a superior estimate of the experimental error at the expense of considerably increased experimental effort. Figure 4 illustrates the idea behind replication. Each set of treatment and noise combinations is run multiple times to produce multiple values of the characteristic response for each treatment condition. Since these values are collected under nominally identical conditions, the difference between them provides an excellent estimate of the experimental error. It is important to note that these multiple measurements must be the result of truly independent experimental runs performed in random order, not multiple measurements on the same run and not sequential measurements made without performing other runs in the interim. Hence the experimental effort scales directly with the number of replications performed.

Replication needs to be clearly distinguished from the collection of data under different noise conditions (described in Chap. 5). In collecting replicates, we attempt to match all of the experimental conditions (both treatment and noise) as closely as possible in order to gain a measure of the pure experimental error. The data collected under different noise conditions, on the other hand, are deliberately collected under different conditions in order to test their influence on the response. Replication is also

TC	Columns			Replicates					
				$k = 1$			$k = 2$		
				NC		$\alpha_{i,1}$	NC		$\alpha_{i,2}$
	A	B	AxB	j=1	j=2		j=1	j=2	
1	-1	-1	+1	$y_{1,1,1}$	$y_{1,2,1}$	$\alpha_{1,1}$	$y_{1,1,2}$	$y_{1,2,2}$	$\alpha_{1,2}$
2	-1	+1	-1	$y_{2,1,1}$	$y_{2,2,1}$	$\alpha_{2,1}$	$y_{2,1,2}$	$y_{2,2,2}$	$\alpha_{2,2}$
3	+1	-1	-1	$y_{3,1,1}$	$y_{3,2,1}$	$\alpha_{3,1}$	$y_{3,1,2}$	$y_{3,2,2}$	$\alpha_{3,2}$
4	+1	+1	+1	$y_{4,1,1}$	$y_{4,2,1}$	$\alpha_{4,1}$	$y_{4,1,2}$	$y_{4,2,2}$	$\alpha_{4,2}$

FIGURE 4 Array design (4 TC Factorial Array) with replication.

Factor	Level	
	-1	+1
A) Launch height	18"	46"
B) launch angle	30°	60°

FIGURE 5 Control factor levels for catapult experiment.

different from the "repeated measurements" strategy for noise. Making repeated measurements during a single experimental run does not capture the effect of random variations occurring from run to run and therefore is not a good measurement of the experimental error. This is why the need to measure each replicate during an independent experimental run was emphasized above.

Determining the experimental error from a replicated experiment can become complicated (and cumbersome). Fortunately, most commercial software packages will do the required computations. Such packages often also allow for additional tests to be performed on the data to further validate the analysis. For a simple design with a uniform number of replicates per treatment condition, such as that shown in Figs. 5–7, an estimate of the error for use in ANOVA can generally be obtained by a process of subtraction. To illustrate use of replication, therefore, we will analyze these experimental results.

Catapult: In this experiment, we are interested in determining the effects of two control factors and their interaction on the range of a small catapult. The catapult needs to handle projectiles with a variety of different added weights, so two different weights (extremes) were chosen as noise conditions. Figures 5 and 6 show the control factor and noise factor/condition levels, respectively. Increasing height (factor A) is obviously

Factor	Noise condition	
	1	2
Projectile weight	low	high

FIGURE 6 Noise condition specification for catapult experiment.

TC	Columns			Replicates					
				k = 1			k = 2		
				NC		$\alpha_{1,1}$	NC		$\alpha_{1,2}$
	A	B	AxB	j=1	j=2		j=1	j=2	
1	-1	-1	+1	81	77	79	93	71	82
2	-1	+1	-1	59	63	61	79	77	78
3	+1	-1	-1	115	107	111	125	109	117
4	+1	+1	+1	95	81	88	92	78	85

FIGURE 7 Array design for replicated catapult experiment, with measured responses and resultant characteristic responses shown.

expected to increase the range, but the role of the launch angle (factor B) is not as clear since neither air resistance nor shifting of the weight in the projectile is considered in the simple dynamics model [Eq. (1) in Chap. 4]. The average range for the two different noise conditions was selected as the characteristic response.

Figure 7 shows the raw data collected from the actual experiment, along with the characteristic response calculated from these data. In particular, note that:

$$\alpha_{i,k} = \frac{\sum_{j=1}^{2} y_{i,j,k}}{2}$$

where $y_{i,j,k}$ is the response measured for the ith treatment condition, the jth noise condition, and the kth replication and $\alpha_{i,k}$ is the characteristic response measured for the ith treatment condition and the kth replication. As a specific example:

$$\alpha_{3,1} = \frac{\sum_{j=1}^{2} y_{3,j,1}}{2} = \frac{y_{3,1,1} + y_{3,2,1}}{2} = \frac{115 + 107}{2} = 111$$

Calculations for ANOM and ANOVA are generally similar to those shown in Chap. 3. However, the number of characteristic responses is no longer equal to the number of treatment conditions. It is necessary to include summations over the replications (k) and account for the effect of replication on the number of characteristic responses being averaged. For example, the overall average for this experiment is given by:

$$m^* = \frac{\sum\limits_{i=1}^{n_{TC}}\sum\limits_{k=1}^{n_r}\alpha_{i,k}}{n_{TC}n_r} = \frac{\sum\limits_{i=1}^{4}\sum\limits_{k=1}^{2}\alpha_{i,k}}{4 \times 2} = 87.625$$

where n_{TC} is the number of treatment conditions and n_r is the number of characteristic responses (replications) for each TC.

DOF: There are a total of 8 dof ($=8$ characteristic responses). One dof is expended to calculate the overall average (m^*). Each of the three two-level columns also has 1 dof associated with it. The remaining 4 ($8-1-3$) dof will be associated with the error estimate.

Sum of squares: The total SS for this experiment (not including the contribution of the mean) is given by:

$$\text{Total SS} = \sum\limits_{i=1}^{n_{TC}}\sum\limits_{k=1}^{n_r}(\alpha_{i,k} - m^*)^2 = 2323.875$$

For each column of the experimental array, we again calculate a variance term, expressed as the sum of squares (SS) for the column. To illustrate, consider the calculations for factor B:

$$m_{+1} = \frac{\alpha_{2,1} + \alpha_{2,2} + \alpha_{4,1} + \alpha_{4,2}}{4} = \frac{61 + 78 + 88 + 85}{4} = 78$$

$$m_{-1} = \frac{\alpha_{1,1} + \alpha_{1,2} + \alpha_{3,1} + \alpha_{3,2}}{4} = \frac{79 + 82 + 111 + 117}{4} = 97.25$$

$$\text{SS (factor B)} = n_{-1}(m_{-1} - m^*)^2 + n_{+1}(m_{+1} - m^*)^2$$

$$= 4(97.25 - 87.625)^2 + 4(78 - 87.625)^2 = 741.125$$

where n_{-1} and n_{+1} are again the number of characteristic responses with a level of -1 and $+1$, respectively, and m_{-1} and m_{+1} are the average value of the characteristic response for level -1 and $+1$, respectively. Since all of the factor and interaction effects are associated with columns in this design, the remainder (i.e., the total minus the sum of all of the columns) provides the error estimate.

Figure 8 shows the completed ANOVA table for the catapult experiment. Using $P_{0.10,1,4}=4.55$, we would judge both the height (A) and the

Source	SS	dof	MS	F
A	1275.125	1	1275.125	**29.7**
B	741.125	1	741.125	**17.3**
AxB	136.125	1	136.125	3.2
error est	171.5	4	42.875	-
Total	2323.875	7	-	-

FIGURE 8 ANOVA for replicated catapult experiment.

angle (B) to be statistically significant. The effect of the $A \times B$ interaction is smaller and does not reach the cutoff for significance.

As illustrated by this example, replication is particularly valuable when the experimental error is quite large and the number of treatment conditions is small. With a small number of treatment conditions, the so-called "pseudo-replication" resulting from averaging the results of many rows (treatment conditions) is less effective in dealing with experimental error. Moreover, when the number of treatment conditions is small, the additional effort required for replication is less burdensome.

IV. NORMAL PROBABILITY PLOTS

A. Concept

Normal probability plots represent an alternative approach to judging statistical significance and are applicable to two-level factorial designs such as those given in Appendix A and C.

To understand the concept of normal probability plots, consider the calculation of the "effect" ($\Delta = m_{+1} - m_{-1}$) for a column. If there is no real effect present (i.e., if changing from a level of -1 to $+1$ has no actual effect on the characteristic response), then the calculation of Δ actually involves adding a (large) number of random outcomes (whether a particular α_I is entered into m_{+1} or m_{-1}). If there are a number of columns with no real effects, the distribution of the Δ's from these columns is expected to follow an approximately normal distribution. (Recall the example of the dice from Chap. 3.) Plotting of these effects on a normal probably plot should yield an approximately straight line. On the other hand, Δ's resulting from real effects will not be part of this distribution and may not be on the line. Thus the position of the Δ for a factor or interaction relative to the line provides a way of judging its significance.

B. Procedure

To conduct an analysis based on this approach, the Δ ($= m_{+1} - m_{-1}$) value for each column in the array is calculated. These are then arranged in order and numbered from the most negative to the most positive. For each effect, a corresponding probability value is then calculated using the following equation:

$$P_k = \frac{100\% \times (k - 0.5)}{n_k} \tag{1}$$

where k is the order ranking for the effect ($k = 1$ is most negative, $k = n_k$ is the most positive), P_k is the probability (coordinate) of the kth point, and n_k is the total number of effects. For each effect, the corresponding point (Δ_k, P_k) is plotted on a normal probability graph. A line is then fit through those points that appear to lie close to a straight line. Points with Δ's that are either too negative or too positive to fit on the line are judged significant.

C. Application Example

For an application example, we will return to our adaptation of the rapid thermal processing study [4], which was briefly described in Chap. 4. In Chap. 4, an ANOVA table was drawn by assuming that all of the interactions were zero and using the resulting "empty" columns to provide an error estimate (Fig. 27). This analysis clearly identified three of the factors (A, D, and E) as significant.

Need for Further Analysis: Although there were no obvious signs of trouble in the ANOVA analysis, the inability to examine interactions is a concern. As a further check, the prediction of a simple model including only the factor effects is useful. Figure 9 shows the average values obtained for the three significant factors at the two levels tested. Since the goal of the experiment was to reduce the resistivity, the best settings of these three factors would be $+1$ for A, -1 for D, and -1 for E. Thus the predicted value of the average resistance under the optimum factor settings is:

$$\alpha_{\text{pred}} = m^* + (m^A_{+1} - m^*) + (m^D_{-1} - m^*) + (m^E_{-1} - m^*)$$

$$= 3.16 + (2.25 - 3.16) + (1.45 - 3.16) + (2.00 - 3.16)$$

$$= -0.62 \ \Omega^2$$

Factor	Level	
	-1	+1
A	4.08	2.25
D	1.45	4.88
E	2.00	4.33

Overall average = 3.16

FIGURE 9   ANOM for significant factors included in the original model for the rapid thermal processing example. (Adapted from Ref. 4.)

Prediction of a negative resistance is obviously a sign of trouble with the model. A likely explanation is the existence of antisynergistic interactions among the factors. Use of the normal probability plot technique can help us check this.

Application: Figure 10 shows the effects measured for each of the columns in the experiment, assembled in order from the most negative to the most positive. Also shown is the matching value of P_k [from Eq. (1)]. These values are plotted on a normal probability graph to give the result shown in Fig. 11. Drawing a line through the appropriate date on the graph, we identify six points that are "off the line." All of them correspond to values which are larger (more negative or positive) than the line and are therefore judged significant. Three of these points correspond to the factors identified before (A, D, and E). This is a nice check on the ANOVA results. The other three correspond to interaction terms (A×D, D×E, and B×C). Note that the ANOVA analysis in Chap. 4 was correct in identifying the significant factors. However, it did not attempt to identify significant interactions (assumed they were zero to provide an error estimate). Hence the model developed was incomplete.

The third of the interactions listed above (B×C) occurs between two factors, which are not by themselves significant, a potential sign of "trouble." Checking the confounding pattern for this design (Fig. 26 in Chap. 4), it is found that B×C is confounded with a three-factor interaction, A×D×E, that involves the strong factors. It is impossible to determine from the experimental results which of the two interactions is responsible for the observed effect. (This is the meaning of confounding.)

Revised Model: Leaving aside the ambiguous BC/ADE interaction, a revised model can be formulated by adding terms for the other two interactions. Figure 12 shows the required ANOM data for the factors and

order (k)	Factor/interaction	Δ_k	P_k
1	A	-1.83	3.3
2	AD	-1.02	10
3	AB	-0.05	16.7
4	CD	-0.04	23.3
5	BD	-0.02	30
6	AC	+0.01	36.7
7	C	+0.10	43.3
8	AE	+0.12	50
9	CE	+0.14	56.7
10	B	+0.20	63.3
11	BE	+0.22	70
12	BC	+0.90	76.7
13	DE	+1.34	83.3
14	E	+2.33	90
15	D	+3.43	96.7

FIGURE 10 Data for generating a normal probability plot for the rapid thermal processing example.

interactions. The interactions are antisynergistic, but in the current case, the interactions will not change the recommended settings. (If needed this could be verified by testing different factor combinations in the model.) The best levels of the three factors remain $+1$ for A, -1 for D, and -1 for E. The corresponding levels for the interactions are determined from the factor levels. Hence for $A \times D$, the level is $-1 (= +1 \times -1)$. For $D \times E$, the level is $+1 (-1 \times -1)$. This gives a revised estimate of:

$$\alpha_{\text{pred}} = m^* + (m^A_{+1} - m^*) + (m^D_{-1} - m^*) + (m^E_{-1} - m^*)$$
$$+ (m^{A \times D}_{-1} - m^*) + (m^{D \times E}_{+1} - m^*)$$

$$= 3.16 + (2.25 - 3.16) + (1.45 - 3.16) + (2.00 - 3.16)$$
$$+ (3.68 - 3.16) + (3.83 - 3.16)$$

$$= +0.57 \ \Omega^2$$

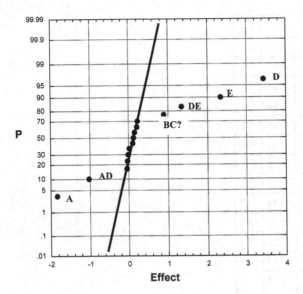

FIGURE 11 Normal probability plot for rapid thermal processing example. (Adapted from Ref. 4.)

Factor/	Level	
Interaction	-1	+1
A	4.08	2.25
D	1.45	4.88
E	2.00	4.33
AD	3.68	2.65
DE	2.49	3.83

Overall average = 3.16

FIGURE 12 ANOM for factors and interactions included in the revised model for the rapid thermal processing example.

This value is still a little lower than the expected minimum (obtained with full film conversion) but is approaching the correct range. As a next step, the possible effects of adding an $A \times D \times E$ interaction term to the model might be examined.

Applicability This technique requires that a number of the effects (columns) are insignificant (i.e., that their values are dominated by random, experimental error). This allows us to draw the straight line used in judging significance. Therefore it is expected to work best with full factorial or high-resolution designs where we believe many of the interactions are near zero. It is most useful in screening a few significant effects (factors and interactions) from a "background" of others. (Having more effects to plot, i.e., using a larger array, also makes establishing the line easier.) In general, the technique should not be used for designs where all or most of the effects are believed to be important since it may be impossible to draw a meaningful line for such cases. In addition, for the simple version we have described here, this technique does sometimes rely on the user's judgment in drawing the line and deciding when a point is off the line.

Homework/Discussion Problems

1) In some respects, pooling of "unused" columns occupies a middle position between the pooling of smallest effects and the use of replication. Discuss.

2) One sign of trouble in a designed experiment is an unexpectedly large sum of squares for the error. Explain how your interpretation of a large error term might be different if the error was estimated by "pooling" of unused columns as opposed to the use of replication.

3) The effects of four control factors on the lifetime of watch batteries was examined using an 8 TC Factorial Array design, with the recommended factor placement to give a resolution IV design. The lifetime (response) was measured under varying environmental (noise) conditions and the average value was used as the characteristic response. Figure P.1 lists the average value (in days) for the -1 and $+1$ levels for each column (three- and four-factor interactions are assumed zero).

   a) Use a normal probability plot to evaluate the significance of the various factors and interactions on the average lifetime.

   b) Use the results from (a) to identify the "best" factor settings and to estimate the average lifetime that would be obtained under these settings.

Factor, Interaction	m_{-1}	m_{+1}
A	675	525
B	598	602
C	584	616
D	640	560
AxB, CxD	590	610
AxC, BxD	604	596
BxC, AxD	550	650

FIGURE P.1 Data for Problem 3.

    c) An eight-treatment condition array is the minimum size array for which use of a normal probability plot is practical and a minimum size of 16 is often recommended. Explain why.

    4) Reanalyze the results of the fluid delivery study (Fig. 8 in Chap. 1) using "pooling of smallest effects." Compare this approach with the original analysis. What additional assumption is necessary to meaningfully analyze the results this way? Do you think this assumption is reasonable in this case?

    5) Use a normal probability plot to reanalyze the results presented in Fig. 1. Is this analysis useful in understanding these results? Discuss.

    6) Reanalyze the results of the fluid delivery study (Fig. 8 in Chap. 1) using a normal probability plot. Are these results consistent with those of the original analysis? Does this new analysis add anything to your understanding?

Appendix A

Two-Level Factorial Arrays

4 TC
8 TC
16 TC
32 TC

4 TC Factorial Array

		Array		
	TC	Columns		
		1	2	3
	1	-1	-1	+1
	2	-1	+1	-1
	3	+1	-1	-1
	4	+1	+1	+1

# of factors	Resolution				
2	Full	Recommended factor placement		A	B
3	III	Recommended factor placement		A	B	C

Notes

2-factors with recommended factor placement

This is a full factorial design, with no confounding of factors and interactions.

Column 1	A
Column 2	B
Column 3	AxB

3-factors with recommended factor placement

This is a resolution III design, with confounding of factors and 2-factor interactions.

Column 1	A, BxC
Column 2	B, AxC
Column 3	C, AxB

Interactions

The interaction between any two of the columns is located in the third column.

8 TC Factorial Array

TC	Array Columns						
	1	2	3	4	5	6	7
1	-1	-1	-1	-1	+1	+1	+1
2	-1	-1	+1	+1	-1	-1	+1
3	-1	+1	-1	+1	-1	+1	-1
4	-1	+1	+1	-1	+1	-1	-1
5	+1	-1	-1	+1	+1	-1	-1
6	+1	-1	+1	-1	-1	+1	-1
7	+1	+1	-1	-1	-1	-1	+1
8	+1	+1	+1	+1	+1	+1	+1

# of factors	Resolution								
3	Full	Factor placement		A	B	C			
4	IV	Factor placement		A	B	C	D		
5-7	III	Factor placement		A	B	C	D	E	F	G

Specifications

3-factors with recommended factor placement
   This is a full factorial design, with no confounding of factors and interactions.

Column	
1	A
2	B
3	C
4	AxBxC
5	BxC
6	AxC
7	AxB

8 TC Factorial Array

4-factors with recommended factor placement

This is a resolution IV design, with confounding of factors and 3-factor interactions.

Column	
1	A, BxCxD
2	B, AxCxD
3	C, AxBxD
4	D, AxBxC
5	BxC, AxD
6	AxC BxD
7	AxB, CxD

5–7 factors

These are resolution III designs, with confounding of factors and 2-factor interactions. The detailed confounding pattern depends on the number of factors and the factor placements.

Interactions

The interaction column for any two columns will have levels corresponding to the product of the levels of the two columns and may be found by inspection. Alternatively the following table may be used:

		Column						
		1	2	3	4	5	6	7
C	1	-						
o	2	7	-					
l	3	6	5	-				
u	4	5	6	7	-			
m	5	4	3	2	1	-		
n	6	3	4	1	2	7	-	
	7	2	1	4	3	6	5	-

16 TC Factorial Array

TC	Array Columns														
	1	2	3	4	5	6	7	8	9	10	11	12	13	14	15
1	-1	-1	-1	-1	+1	-1	-1	+1	-1	+1	+1	-1	+1	+1	+1
2	-1	-1	-1	+1	-1	+1	+1	-1	+1	-1	-1	-1	+1	+1	+1
3	-1	-1	+1	-1	-1	+1	+1	-1	-1	+1	+1	+1	-1	-1	+1
4	-1	-1	+1	+1	+1	-1	-1	+1	+1	-1	-1	+1	-1	-1	+1
5	-1	+1	-1	-1	-1	+1	-1	+1	+1	-1	+1	+1	-1	+1	-1
6	-1	+1	-1	+1	+1	-1	+1	-1	-1	+1	-1	+1	-1	+1	-1
7	-1	+1	+1	-1	+1	-1	+1	-1	+1	-1	+1	-1	+1	-1	-1
8	-1	+1	+1	+1	-1	+1	-1	+1	-1	+1	-1	-1	+1	-1	-1
9	+1	-1	-1	-1	-1	-1	+1	+1	+1	+1	-1	+1	+1	-1	-1
10	+1	-1	-1	+1	+1	+1	-1	-1	-1	-1	+1	+1	+1	-1	-1
11	+1	-1	+1	-1	+1	+1	-1	-1	+1	+1	-1	-1	-1	+1	-1
12	+1	-1	+1	+1	-1	-1	+1	+1	-1	-1	+1	-1	-1	+1	-1
13	+1	+1	-1	-1	+1	+1	+1	+1	-1	-1	-1	-1	-1	-1	+1
14	+1	+1	-1	+1	-1	-1	-1	-1	+1	+1	+1	-1	-1	-1	+1
15	+1	+1	+1	-1	-1	-1	-1	-1	-1	-1	-1	+1	+1	+1	+1
16	+1	+1	+1	+1	+1	+1	+1	+1	+1	+1	+1	+1	+1	+1	+1

# of factors	Resolution	1	2	3	4	5	6	7	8	9	10	11	12	13	14	15
4	FF	A	B	C	D											
5	V	A	B	C	D	E										
6-8	IV	A	B	C	D		E	F		G			H			
9-15	III	A	B	C	D	E	F	G	H	I	J	K	L	M	N	O

16 TC Factorial Array

Specifications

4-factors with recommended factor placement

    This is a full factorial design, with no confounding of factors and interactions.

Column	
1	A
2	B
3	C
4	D
5	AxBxCxD
6	BxCxD
7	AxCxD
8	CxD
9	AxBxD
10	BxD
11	AxD
12	AxBxC
13	BxC
14	AxC
15	AxB

5-factors with recommended factor placement

    This is a resolution V design, with confounding of factors and 4-factor interactions.

Column	
1	A, BxCxDxE
2	B, AxCxDxE
3	C, AxBxDxE
4	D, AxBxCxE
5	E, AxBxCxD
6	AxE, BxCxD
7	BxE, AxCxD
8	CxD, AxBxE
9	CxE, AxBxD
10	BxD, AxCxE
11	AxD, BxCxE
12	DxE, AxBxC
13	BxC, AxDxE
14	AxC, BxDxE
15	AxB, CxDxE

16 TC Factorial Array

6–8 factors

These are resolution IV designs, with confounding of factors and 3-factor interactions. The detailed confounding pattern depends on the number of factors and the factor placements. It may be developed with aid of the interaction table.

9–15 factors

These are resolution III designs, with confounding of factors and 2-factor interactions. The detailed confounding pattern depends on the number of factors and the factor placements. It may be developed with the aid of the interaction table.

Interactions

The interaction column for any two columns will have levels corresponding to the product of the levels of the two columns and may be found by inspection. Alternatively the following table may be used:

		Column														
		1	2	3	4	5	6	7	8	9	10	11	12	13	14	15
	1															
	2	15	-													
	3	14	13	-												
	4	11	10	8	-											
C	5	6	7	9	12	-										
o	6	5	8	10	13	1	-									
l	7	8	5	11	14	2	15	-								
u	8	7	6	4	3	15	2	1	-							
m	9	10	11	5	15	3	14	13	12	-						
n	10	9	4	6	2	14	3	12	13	1	-					
	11	4	9	7	1	13	12	3	14	2	15	-				
	12	13	14	15	5	4	11	10	9	8	7	6	-			
	13	12	3	2	6	11	4	9	10	7	8	5	1	-		
	14	3	12	1	7	10	9	4	11	6	5	8	2	15	-	
	15	2	1	12	9	8	7	6	5	4	11	10	3	14	13	-

32 TC Factorial Array

Array (Columns 1-16, remaining columns on next page)

TC	1	2	3	4	5	6	7	8	9	10	11	12	13	14	15	16
1	-1	-1	-1	-1	-1	-1	+1	+1	-1	+1	-1	-1	+1	+1	-1	-1
2	-1	-1	-1	-1	+1	+1	-1	-1	+1	-1	+1	+1	-1	-1	+1	+1
3	-1	-1	-1	+1	-1	+1	-1	-1	+1	-1	+1	+1	-1	+1	-1	-1
4	-1	-1	-1	+1	+1	-1	+1	+1	-1	+1	-1	-1	+1	-1	+1	+1
5	-1	-1	+1	-1	-1	+1	-1	-1	+1	+1	-1	-1	+1	-1	+1	+1
6	-1	-1	+1	-1	+1	-1	+1	+1	-1	-1	+1	+1	-1	+1	-1	-1
7	-1	-1	+1	+1	-1	-1	+1	+1	-1	-1	+1	+1	-1	-1	+1	+1
8	-1	-1	+1	+1	+1	+1	-1	-1	+1	+1	-1	-1	+1	+1	-1	-1
9	-1	+1	-1	-1	-1	+1	-1	+1	-1	-1	+1	-1	+1	-1	+1	-1
10	-1	+1	-1	-1	+1	-1	+1	-1	+1	+1	-1	+1	-1	+1	-1	+1
11	-1	+1	-1	+1	-1	-1	+1	-1	+1	+1	-1	+1	-1	-1	+1	-1
12	-1	+1	-1	+1	+1	+1	-1	+1	-1	-1	+1	-1	+1	+1	-1	+1
13	-1	+1	+1	-1	-1	-1	+1	-1	+1	-1	+1	-1	+1	+1	-1	+1
14	-1	+1	+1	-1	+1	+1	-1	+1	-1	+1	-1	+1	-1	-1	+1	-1
15	-1	+1	+1	+1	-1	+1	-1	+1	-1	+1	-1	+1	-1	+1	-1	+1
16	-1	+1	+1	+1	+1	-1	+1	-1	+1	-1	+1	-1	+1	-1	+1	-1
17	+1	-1	-1	-1	-1	+1	+1	-1	-1	-1	-1	+1	+1	-1	-1	+1
18	+1	-1	-1	-1	+1	-1	-1	+1	+1	+1	+1	-1	-1	+1	+1	-1
19	+1	-1	-1	+1	-1	-1	-1	+1	+1	+1	+1	-1	-1	-1	-1	+1
20	+1	-1	-1	+1	+1	+1	+1	-1	-1	-1	-1	+1	+1	+1	+1	-1
21	+1	-1	+1	-1	-1	-1	-1	+1	+1	-1	-1	+1	+1	+1	+1	-1
22	+1	-1	+1	-1	+1	+1	+1	-1	-1	+1	+1	-1	-1	-1	-1	+1
23	+1	-1	+1	+1	-1	+1	+1	-1	-1	+1	+1	-1	-1	+1	+1	-1
24	+1	-1	+1	+1	+1	-1	-1	+1	+1	-1	-1	+1	+1	-1	-1	+1
25	+1	+1	-1	-1	-1	-1	-1	-1	-1	+1	+1	+1	+1	+1	+1	+1
26	+1	+1	-1	-1	+1	+1	+1	+1	+1	-1	-1	-1	-1	-1	-1	-1
27	+1	+1	-1	+1	-1	+1	+1	+1	+1	-1	-1	-1	-1	+1	+1	+1
28	+1	+1	-1	+1	+1	-1	-1	-1	-1	+1	+1	+1	+1	-1	-1	-1
29	+1	+1	+1	-1	-1	+1	+1	+1	+1	+1	+1	+1	+1	-1	-1	-1
30	+1	+1	+1	-1	+1	-1	-1	-1	-1	-1	-1	-1	-1	+1	+1	+1
31	+1	+1	+1	+1	-1	-1	-1	-1	-1	-1	-1	-1	-1	-1	-1	-1
32	+1	+1	+1	+1	+1	+1	+1	+1	+1	+1	+1	+1	+1	+1	+1	+1

32 TC Factorial Array

Array (Columns 17-32, remaining columns on previous page)

TC	Columns														
	17	18	19	20	21	22	23	24	25	26	27	28	29	30	31
1	+1	-1	+1	+1	+1	-1	-1	+1	-1	+1	+1	-1	+1	+1	+1
2	-1	+1	-1	-1	+1	-1	-1	+1	-1	+1	+1	-1	+1	+1	+1
3	+1	-1	+1	+1	-1	+1	+1	-1	+1	-1	-1	-1	+1	+1	+1
4	-1	+1	-1	-1	-1	+1	+1	-1	+1	-1	-1	-1	+1	+1	+1
5	-1	-1	+1	+1	-1	+1	+1	-1	-1	+1	+1	+1	-1	-1	+1
6	+1	+1	-1	-1	-1	+1	+1	-1	-1	+1	+1	+1	-1	-1	+1
7	-1	-1	+1	+1	+1	-1	-1	+1	+1	-1	-1	+1	-1	-1	+1
8	+1	+1	-1	-1	+1	-1	-1	+1	+1	-1	-1	+1	-1	-1	+1
9	+1	+1	-1	+1	-1	+1	-1	+1	+1	-1	+1	+1	-1	+1	-1
10	-1	-1	+1	-1	-1	+1	-1	+1	+1	-1	+1	+1	-1	+1	-1
11	+1	+1	-1	+1	+1	-1	+1	-1	-1	+1	-1	+1	-1	+1	-1
12	-1	-1	+1	-1	+1	-1	+1	-1	-1	+1	-1	+1	-1	+1	-1
13	-1	+1	-1	+1	+1	-1	+1	-1	+1	-1	+1	-1	+1	-1	-1
14	+1	-1	+1	-1	+1	-1	+1	-1	+1	-1	+1	-1	+1	-1	-1
15	-1	+1	-1	+1	-1	+1	-1	+1	-1	+1	-1	-1	+1	-1	-1
16	+1	-1	+1	-1	-1	+1	-1	+1	-1	+1	-1	-1	+1	-1	-1
17	+1	+1	+1	-1	-1	-1	+1	+1	+1	+1	-1	+1	+1	-1	-1
18	-1	-1	-1	+1	-1	-1	+1	+1	+1	+1	-1	+1	+1	-1	-1
19	+1	+1	+1	-1	+1	+1	-1	-1	-1	-1	+1	+1	+1	-1	-1
20	-1	-1	-1	+1	+1	+1	-1	-1	-1	-1	+1	+1	+1	-1	-1
21	-1	+1	+1	-1	+1	+1	-1	-1	+1	+1	-1	-1	-1	+1	-1
22	+1	-1	-1	+1	+1	+1	-1	-1	+1	+1	-1	-1	-1	+1	-1
23	-1	+1	+1	-1	-1	-1	+1	+1	-1	-1	+1	-1	-1	+1	-1
24	+1	-1	-1	+1	-1	-1	+1	+1	-1	-1	+1	-1	-1	+1	-1
25	+1	-1	-1	-1	+1	+1	+1	+1	-1	-1	-1	-1	-1	-1	+1
26	-1	+1	+1	+1	+1	+1	+1	+1	-1	-1	-1	-1	-1	-1	+1
27	+1	-1	-1	-1	-1	-1	-1	-1	+1	+1	+1	-1	-1	-1	+1
28	-1	+1	+1	+1	-1	-1	-1	-1	+1	+1	+1	-1	-1	-1	+1
29	-1	-1	-1	-1	-1	-1	-1	-1	-1	-1	-1	+1	+1	+1	+1
30	+1	+1	+1	+1	-1	-1	-1	-1	-1	-1	-1	+1	+1	+1	+1
31	-1	-1	-1	-1	+1	+1	+1	+1	+1	+1	+1	+1	+1	+1	+1
32	+1	+1	+1	+1	+1	+1	+1	+1	+1	+1	+1	+1	+1	+1	+1

32 TC Factorial Array

5-factors

Recommended factor placement is in columns 1, 2, 3, 4, and 5. This produces a full factorial design. Factor/Interaction locations for this assignment are summarized below.

Column		Column	
1	A	17	CxE
2	B	18	AxBxE
3	C	19	BxE
4	D	20	AxE
5	E	21	AxBxCxD
6	AxBxCxDxE	22	BxCxD
7	BxCxDxE	23	AxCxD
8	AxCxDxE	24	CxD
9	CxDxE	25	AxBxD
10	AxBxDxE	26	BxD
11	BxDxE	27	AxD
12	AxDxE	28	AxBxC
13	DxE	29	BxC
14	AxBxCxE	30	AxC
15	BxCxE	31	AxB
16	AxCxE		

32 TC Factorial Array

6-factors

Recommended factor placement is in columns 1, 2, 3, 4, 5 and 6. This produces a resolution VI design. Factor/Interaction locations for this assignment are summarized below. Only 2 and 3 factor interactions are shown.

Column		Column	
1	A	17	CxE
2	B	18	AxBxE, CxDxF
3	C	19	BxE
4	D	20	AxE
5	E	21	ExF
6	F	22	BxCxD, AxExF
7	AxF	23	AxCxD, BxExF
8	BxF	24	CxD
9	CxDxE, AxBxF	25	AxBxD, CxExF
10	CxF	26	BxD
11	BxDxE, AxCxF	27	AxD
12	AxDxE, BxCxF	28	AxBxC, DxExF
13	DxE	29	BxC
14	DxF	30	AxC
15	BxCxE, AxDxF	31	AxB
16	AxCxE, BxDxF		

7–16 factors

Recommended factor placement is in columns 1, 2, 3, 4, 5, 6, 9, 11, 12, 15, 16, 18, 22, 23, 25, and 28. These are resolution IV designs, with confounding of factors and 3-factor interactions. The detailed confounding pattern depends on the number of factors and the factor placements. It may be developed with aid of the interaction table.

17–31 factors

These are resolution III designs, with confounding of factors and 2-factor interactions. The detailed confounding pattern depends on the number of factors and the factor placements. It may be developed with aid of the interaction table.

32 TC Factorial Array

Interactions

The interaction column for any two columns will have levels corresponding to the product of the levels of the two columns and may be found by inspection. Alternatively the following table may be used:

Array (Columns 1-16, remaining columns on next page)

		1	2	3	4	5	6	7	8	9	10	11	12	13	14	15	16
	1	-															
	2	31	-														
	3	30	29	-													
	4	27	26	24	-												
	5	20	19	17	13	-											
	6	7	8	10	14	21	-										
	7	6	9	11	15	22	1	-									
	8	9	6	12	16	23	2	31	-								
	9	8	7	13	17	24	31	2	1	-							
	10	11	12	6	18	25	3	30	29	28	-						
	11	10	13	7	19	26	30	3	28	29	1	-					
	12	13	10	8	20	27	29	28	3	30	2	31	-				
	13	12	11	9	5	4	28	29	30	3	31	2	1	-			
C	14	15	16	18	6	28	4	27	26	25	24	23	22	21	-		
o	15	14	17	19	7	29	27	4	25	26	23	24	21	22	1	-	
l	16	17	14	20	8	30	26	25	4	27	22	21	24	23	2	31	-
u	17	16	15	5	9	3	25	26	27	4	21	22	23	24	31	2	1
m	18	19	20	14	10	31	24	23	22	21	4	27	26	25	3	30	29
n	19	18	5	15	11	2	23	24	21	22	27	4	25	26	30	3	28
	20	5	18	16	12	1	22	21	24	23	26	25	4	27	29	28	3
	21	22	23	25	28	6	5	20	19	18	17	16	15	14	13	12	11
	22	21	24	26	29	7	20	5	18	19	16	17	14	15	12	13	10
	23	24	21	27	30	8	19	18	5	20	15	14	17	16	11	10	13
	24	23	22	4	3	9	18	19	20	5	14	15	16	17	10	11	12
	25	26	27	21	31	10	17	16	15	14	5	20	19	18	9	8	7
	26	25	4	22	2	11	16	17	14	15	20	5	18	19	8	9	6
	27	4	25	23	1	12	15	14	17	16	19	18	5	20	7	6	9
	28	29	30	31	21	14	13	12	11	10	9	8	7	6	5	20	19
	29	28	3	2	22	15	12	13	10	11	8	9	6	7	20	5	18
	30	3	28	1	23	16	11	10	13	12	7	6	9	8	19	18	5
	31	2	1	28	25	18	9	8	7	6	13	12	11	10	17	16	15

32 TC Factorial Array

Array (Columns 17-32, remaining columns on previous page)

Column	17	18	19	20	21	22	23	24	25	26	27	28	29	30	31
1															
2															
3															
4															
5															
6															
7															
8															
9															
10															
11															
12															
13															
14															
15															
16															
17	-														
18	28	-													
19	29	1	-												
20	30	2	31	-											
21	10	9	8	7	-										
22	11	8	9	6	1	-									
23	12	7	6	9	2	31	-								
24	13	6	7	8	31	2	1	-							
25	6	13	12	11	3	30	29	28	-						
26	7	12	13	10	30	3	28	29	1	-					
27	8	11	10	13	29	28	3	30	2	31	-				
28	18	17	16	15	4	27	26	25	24	23	22	-			
29	19	16	17	14	27	4	25	26	23	24	21	1	-		
30	20	15	14	17	26	25	4	27	22	21	24	2	31	-	
31	14	5	20	19	24	23	22	21	4	27	26	3	30	29	-

Appendix B

Three-Level Factorial Arrays

9 TC
27 TC

9 TC Factorial Array

			Array				
			TC	Columns			
				1	2	3	4
			1	-1	-1	-1	-1
			2	-1	0	+1	+1
			3	-1	+1	0	0
			4	0	-1	+1	0
			5	0	0	0	-1
			6	0	+1	-1	+1
			7	+1	-1	0	+1
			8	+1	0	-1	0
			9	+1	+1	+1	-1
# of factors	Resolution						
2	Full	Factor placement		A	B		
4	III	Factor placement		A	B	C	D

Notes

Two factors with recommended factor placement: This is a full factorial design, with no confounding of factors and interactions.

Column 1	A
Column 2	B
Column 3	AxB
Column 4	AxB

Three to four factors: These are resolution III designs, with confounding of factors and two-factor interactions.

Interactions

The interaction between any two of the columns is located in the other two columns.

27 TC Factorial Array

Array

TC	1	2	3	4	5	6	7	8	9	10	11	12	13
					Columns								
1	-1	-1	-1	-1	-1	-1	-1	-1	-1	-1	-1	-1	-1
2	-1	-1	0	+1	+1	+1	+1	+1	+1	+1	+1	-1	-1
3	-1	-1	+1	0	0	0	0	0	0	0	0	-1	-1
4	-1	0	-1	+1	+1	+1	0	0	0	-1	-1	+1	+1
5	-1	0	0	0	0	0	-1	-1	-1	+1	+1	+1	+1
6	-1	0	+1	-1	-1	-1	+1	+1	+1	0	0	+1	+1
7	-1	+1	-1	0	0	0	+1	+1	+1	-1	-1	0	0
8	-1	+1	0	-1	-1	-1	0	0	0	+1	+1	0	0
9	-1	+1	+1	+1	+1	+1	-1	-1	-1	0	0	0	0
10	0	-1	-1	+1	0	-1	+1	0	-1	+1	0	+1	0
11	0	-1	0	0	-1	+1	0	-1	+1	0	-1	+1	0
12	0	-1	+1	-1	+1	0	-1	+1	0	-1	+1	+1	0
13	0	0	-1	0	-1	+1	-1	+1	0	+1	0	0	-1
14	0	0	0	-1	+1	0	+1	0	-1	0	-1	0	-1
15	0	0	+1	+1	0	-1	0	-1	+1	-1	+1	0	-1
16	0	+1	-1	-1	+1	0	0	-1	+1	+1	0	-1	+1
17	0	+1	0	+1	0	-1	-1	+1	0	0	-1	-1	+1
18	0	+1	+1	0	-1	+1	+1	0	-1	-1	+1	-1	+1
19	+1	-1	-1	0	+1	-1	0	+1	-1	0	+1	0	+1
20	+1	-1	0	-1	0	+1	-1	0	+1	-1	0	0	+1
21	+1	-1	+1	+1	-1	0	+1	-1	0	+1	-1	0	+1
22	+1	0	-1	-1	0	+1	+1	-1	0	0	+1	-1	0
23	+1	0	0	+1	-1	0	0	+1	-1	-1	0	-1	0
24	+1	0	+1	0	+1	-1	-1	0	+1	+1	-1	-1	0
25	+1	+1	-1	+1	-1	0	-1	0	+1	0	+1	+1	-1
26	+1	+1	0	0	+1	-1	+1	-1	0	-1	0	+1	-1
27	+1	+1	+1	-1	0	+1	0	+1	-1	+1	-1	+1	-1

# of factors	Resolution	1	2	3	4	5	6	7	8	9	10	11	12	13
3	Full	A	B	C										
4	IV	A	B	C	D									
5-13	III	A	B	C	D	E	F	G	H	I	J	K	L	M

27 TC Factorial Array
Specifications

Three factors with recommended factor placement: This is a full factorial design, with no confounding of factors and interactions.

Column	
1	A
2	B
3	C
4	AxBxC
5	AxBxC
6	BxC
7	AxBxC
8	AxBxC
9	BxC
10	AxC
11	AxC
12	AxB
13	AxB

Four factors with recommended factor placement: This is a resolution IV design, with confounding of factors and three-factor interactions. The following table shows only factors and two-factor interactions.

Column	
1	A
2	B
3	C
4	D
5	AxD
6	AxD, BxC
7	BxD
8	CxD
9	BxC
10	BxD, AxC
11	AxC
12	CxD, AxB
13	AxB

27 TC Factorial Array

Five to thirteen factors: These are resolution III designs, with confounding of factors and two-factor interactions. The detailed confounding
pattern depends on the number of factors and factor placements. It may be
developed with the aid of the interaction table.

Interactions

The interaction for any two columns occupies two columns.

		1	2	3	4	5	Column 6	7	8	9	10	11	12	13
	1	-												
	2	12, 13	-											
	3	10, 11	6, 9	-										
	4	5, 6	7, 10	8, 12	-									
C	5	4, 6	8, 11	7, 13	1, 6	-								
o	6	4, 5	3, 9	2, 9	1, 5	1, 4	-							
l	7	8, 9	4, 10	5, 13	2, 10	3, 13	11, 12	-						
u	8	7, 9	5, 11	4, 12	3, 12	2, 11	10, 13	1, 9	-					
m	9	7, 8	3, 6	2, 6	11, 13	10, 12	2, 3	1, 8	1, 7	-				
n	10	3, 11	4, 7	1, 11	2, 7	9, 12	8, 13	2, 4	6, 13	5, 12	-			
	11	3, 10	5, 8	1, 10	9, 13	2, 8	7, 12	6, 12	2, 5	4, 13	1, 3	-		
	12	2, 13	1, 13	4, 8	3, 8	9, 10	7, 11	6, 11	3, 4	5, 10	5, 9	6, 7	-	
	13	2, 12	1, 12	5, 7	9, 11	3, 7	8, 10	3, 5	6, 10	4, 11	6, 8	4, 9	1, 2	-

Appendix C

Screening Arrays

12 TC
18 TC
24 TC

12 TC Screening Array (Plackett–Burman type)

Array

TC	Columns										
	1	2	3	4	5	6	7	8	9	10	11
1	+1	+1	+1	+1	+1	+1	+1	+1	+1	+1	+1
2	-1	+1	-1	+1	+1	+1	-1	-1	-1	+1	-1
3	-1	-1	+1	-1	+1	+1	+1	-1	-1	-1	+1
4	+1	-1	-1	+1	-1	+1	+1	+1	-1	-1	-1
5	-1	+1	-1	-1	+1	-1	+1	+1	+1	-1	-1
6	-1	-1	+1	-1	-1	+1	-1	+1	+1	+1	-1
7	-1	-1	-1	+1	-1	-1	+1	-1	+1	+1	+1
8	+1	-1	-1	-1	+1	-1	-1	+1	-1	+1	+1
9	+1	+1	-1	-1	-1	+1	-1	-1	+1	-1	+1
10	+1	+1	+1	-1	-1	-1	+1	-1	-1	+1	-1
11	-1	+1	+1	+1	-1	-1	-1	+1	-1	-1	+1
12	+1	-1	+1	+1	+1	-1	-1	-1	+1	-1	-1

18 TC Screening Array (L18 type)

Array

TC	Columns							
	1	2	3	4	5	6	7	8
1	-1	-1	-1	-1	-1	-1	-1	-1
2	-1	-1	0	0	+1	0	0	0
3	-1	-1	+1	+1	0	+1	+1	+1
4	-1	0	-1	0	0	0	+1	-1
5	-1	0	0	+1	-1	+1	-1	0
6	-1	0	+1	-1	+1	-1	0	+1
7	-1	+1	-1	+1	+1	0	-1	+1
8	-1	+1	0	-1	0	+1	0	-1
9	-1	+1	+1	0	-1	-1	+1	0
10	+1	-1	-1	+1	0	-1	0	0
11	+1	-1	0	-1	-1	0	+1	+1
12	+1	-1	+1	0	+1	+1	-1	-1
13	+1	0	-1	-1	+1	+1	+1	0
14	+1	0	0	0	0	-1	-1	+1
15	+1	0	+1	+1	-1	0	0	-1
16	+1	+1	-1	0	-1	+1	0	+1
17	+1	+1	0	+1	+1	-1	+1	-1
18	+1	+1	+1	-1	0	0	-1	0

24 TC Factorial Array (Plackett–Burman type)

Array (Columns 1-12, remaining columns on next page)

TC	Columns											
	1	2	3	4	5	6	7	8	9	10	11	12
1	+1	+1	+1	+1	+1	+1	+1	+1	+1	+1	+1	+1
2	-1	+1	+1	+1	+1	-1	+1	-1	+1	+1	-1	-1
3	-1	-1	+1	+1	+1	+1	-1	+1	-1	+1	+1	-1
4	-1	-1	-1	+1	+1	+1	+1	-1	+1	-1	+1	+1
5	-1	-1	-1	-1	+1	+1	+1	+1	-1	+1	-1	+1
6	-1	-1	-1	-1	-1	+1	+1	+1	+1	-1	+1	-1
7	+1	-1	-1	-1	-1	-1	+1	+1	+1	+1	-1	+1
8	-1	+1	-1	-1	-1	-1	-1	+1	+1	+1	+1	-1
9	+1	-1	+1	-1	-1	-1	-1	-1	+1	+1	+1	+1
10	-1	+1	-1	+1	-1	-1	-1	-1	-1	+1	+1	+1
11	-1	-1	+1	-1	+1	-1	-1	-1	-1	-1	+1	+1
12	+1	-1	-1	+1	-1	+1	-1	-1	-1	-1	-1	+1
13	+1	+1	-1	-1	+1	-1	+1	-1	-1	-1	-1	-1
14	-1	+1	+1	-1	-1	+1	-1	+1	-1	-1	-1	-1
15	-1	-1	+1	+1	-1	-1	+1	-1	+1	-1	-1	-1
16	+1	-1	-1	+1	+1	-1	-1	+1	-1	+1	-1	-1
17	+1	+1	-1	-1	+1	+1	-1	-1	+1	-1	+1	-1
18	-1	+1	+1	-1	-1	+1	+1	-1	-1	+1	-1	+1
19	+1	-1	+1	+1	-1	-1	+1	+1	-1	-1	+1	-1
20	-1	+1	-1	+1	+1	-1	-1	+1	+1	-1	-1	+1
21	+1	-1	+1	-1	+1	+1	-1	-1	+1	+1	-1	-1
22	+1	+1	-1	+1	-1	+1	+1	-1	-1	+1	+1	-1
23	+1	+1	+1	-1	+1	-1	+1	+1	-1	-1	+1	+1
24	+1	+1	+1	+1	-1	+1	-1	+1	+1	-1	-1	+1

24 TC Factorial Array

Array (Columns 13-23, remaining columns on previous page)

TC	Columns										
	13	14	15	16	17	18	19	20	21	22	23
1	+1	+1	+1	+1	+1	+1	+1	+1	+1	+1	+1
2	+1	+1	-1	-1	+1	-1	+1	-1	-1	-1	-1
3	-1	+1	+1	-1	-1	+1	-1	+1	-1	-1	-1
4	-1	-1	+1	+1	-1	-1	+1	-1	+1	-1	-1
5	+1	-1	-1	+1	+1	-1	-1	+1	-1	+1	-1
6	+1	+1	-1	-1	+1	+1	-1	-1	+1	-1	+1
7	-1	+1	+1	-1	-1	+1	+1	-1	-1	+1	-1
8	+1	-1	+1	+1	-1	-1	+1	+1	-1	-1	+1
9	-1	+1	-1	+1	+1	-1	-1	+1	+1	-1	-1
10	+1	-1	+1	-1	+1	+1	-1	-1	+1	+1	-1
11	+1	+1	-1	+1	-1	+1	+1	-1	-1	+1	+1
12	+1	+1	+1	-1	+1	-1	+1	+1	-1	-1	+1
13	+1	+1	+1	+1	-1	+1	-1	+1	+1	-1	-1
14	-1	+1	+1	+1	+1	-1	+1	-1	+1	+1	-1
15	-1	-1	+1	+1	+1	+1	-1	+1	-1	+1	+1
16	-1	-1	-1	+1	+1	+1	+1	-1	+1	-1	+1
17	-1	-1	-1	-1	+1	+1	+1	+1	-1	+1	-1
18	-1	-1	-1	-1	-1	+1	+1	+1	+1	-1	+1
19	+1	-1	-1	-1	-1	-1	+1	+1	+1	+1	-1
20	-1	+1	-1	-1	-1	-1	-1	+1	+1	+1	+1
21	+1	-1	+1	-1	-1	-1	-1	-1	+1	+1	+1
22	-1	+1	-1	+1	-1	-1	-1	-1	-1	+1	+1
23	-1	-1	+1	-1	+1	-1	-1	-1	-1	-1	+1
24	+1	-1	-1	+1	-1	+1	-1	-1	-1	-1	-1

Appendix D

Critical Values of F

Critical F for P = 0.90

v_2	1	2	3	4	5	6	7	8
1	39.86	49.50	53.59	55.83	57.24	58.20	58.91	59.44
2	8.53	9.00	9.16	9.24	9.29	9.33	9.35	9.37
3	5.54	5.46	5.39	5.34	5.31	5.28	5.27	5.25
4	4.54	4.32	4.19	4.11	4.05	4.01	3.98	3.95
5	4.06	3.78	3.62	3.52	3.45	3.40	3.37	3.34
6	3.78	3.46	3.29	3.18	3.11	3.05	3.01	2.98
7	3.59	3.26	3.07	2.96	2.88	2.83	2.78	2.75
8	3.46	3.11	2.92	2.81	2.73	2.67	2.62	2.59
9	3.36	3.01	2.81	2.69	2.61	2.55	2.51	2.47
10	3.29	2.92	2.73	2.61	2.52	2.46	2.41	2.38
11	3.23	2.86	2.66	2.54	2.45	2.39	2.34	2.30
12	3.18	2.81	2.61	2.48	2.39	2.33	2.28	2.24
13	3.14	2.76	2.56	2.43	2.35	2.28	2.23	2.20
14	3.10	2.73	2.52	2.39	2.31	2.24	2.19	2.15
15	3.07	2.70	2.49	2.36	2.27	2.21	2.16	2.12
16	3.05	2.67	2.46	2.33	2.24	2.18	2.13	2.09
17	3.03	2.64	2.44	2.31	2.22	2.15	2.10	2.06
18	3.01	2.62	2.42	2.29	2.20	2.13	2.08	2.04
19	2.99	2.61	2.40	2.27	2.18	2.11	2.06	2.02
20	2.97	2.59	2.38	2.25	2.16	2.09	2.04	2.00
25	2.92	2.53	2.32	2.18	2.09	2.02	1.97	1.93
30	2.88	2.49	2.28	2.14	2.05	1.98	1.93	1.88
35	2.85	2.46	2.25	2.11	2.02	1.95	1.90	1.85
∞	2.71	2.30	2.08	1.95	1.85	1.77	1.72	1.67

Column headers span under v_1.

Critical F for P = 0.95

v_2	v_1							
	1	2	3	4	5	6	7	8
1	161.45	199.50	215.71	224.58	230.16	233.99	236.77	238.88
2	18.51	19.00	19.16	19.25	19.30	19.33	19.35	19.37
3	10.13	9.55	9.28	9.12	9.01	8.94	8.89	8.85
4	7.71	6.94	6.59	6.39	6.26	6.16	6.09	6.04
5	6.61	5.79	5.41	5.19	5.05	4.95	4.88	4.82
6	5.99	5.14	4.76	4.53	4.39	4.28	4.21	4.15
7	5.59	4.74	4.35	4.12	3.97	3.87	3.79	3.73
8	5.32	4.46	4.07	3.84	3.69	3.58	3.50	3.44
9	5.12	4.26	3.86	3.63	3.48	3.37	3.29	3.23
10	4.96	4.10	3.71	3.48	3.33	3.22	3.14	3.07
11	4.84	3.98	3.59	3.36	3.20	3.09	3.01	2.95
12	4.75	3.89	3.49	3.26	3.11	3.00	2.91	2.85
13	4.67	3.81	3.41	3.18	3.03	2.92	2.83	2.77
14	4.60	3.74	3.34	3.11	2.96	2.85	2.76	2.70
15	4.54	3.68	3.29	3.06	2.90	2.79	2.71	2.64
16	4.49	3.63	3.24	3.01	2.85	2.74	2.66	2.59
17	4.45	3.59	3.20	2.96	2.81	2.70	2.61	2.55
18	4.41	3.55	3.16	2.93	2.77	2.66	2.58	2.51
19	4.38	3.52	3.13	2.90	2.74	2.63	2.54	2.48
20	4.35	3.49	3.10	2.87	2.71	2.60	2.51	2.45
25	4.24	3.39	2.99	2.76	2.60	2.49	2.40	2.34
30	4.17	3.32	2.92	2.69	2.53	2.42	2.33	2.27
35	4.12	3.27	2.87	2.64	2.49	2.37	2.29	2.22
∞	3.84	3.00	2.60	2.37	2.21	2.10	2.01	1.94

Critical F for P = 0.99

v_2	v_1							
	1	2	3	4	5	6	7	8
1	4052.18	4999.34	5403.53	5624.26	5763.96	5858.95	5928.33	5980.95
2	98.50	99.00	99.16	99.25	99.30	99.33	99.36	99.38
3	34.12	30.82	29.46	28.71	28.24	27.91	27.67	27.49
4	21.20	18.00	16.69	15.98	15.52	15.21	14.98	14.80
5	16.26	13.27	12.06	11.39	10.97	10.67	10.46	10.29
6	13.75	10.92	9.78	9.15	8.75	8.47	8.26	8.10
7	12.25	9.55	8.45	7.85	7.46	7.19	6.99	6.84
8	11.26	8.65	7.59	7.01	6.63	6.37	6.18	6.03
9	10.56	8.02	6.99	6.42	6.06	5.80	5.61	5.47
10	10.04	7.56	6.55	5.99	5.64	5.39	5.20	5.06
11	9.65	7.21	6.22	5.67	5.32	5.07	4.89	4.74
12	9.33	6.93	5.95	5.41	5.06	4.82	4.64	4.50
13	9.07	6.70	5.74	5.21	4.86	4.62	4.44	4.30
14	8.86	6.51	5.56	5.04	4.69	4.46	4.28	4.14
15	8.68	6.36	5.42	4.89	4.56	4.32	4.14	4.00
16	8.53	6.23	5.29	4.77	4.44	4.20	4.03	3.89
17	8.40	6.11	5.19	4.67	4.34	4.10	3.93	3.79
18	8.29	6.01	5.09	4.58	4.25	4.01	3.84	3.71
19	8.18	5.93	5.01	4.50	4.17	3.94	3.77	3.63
20	8.10	5.85	4.94	4.43	4.10	3.87	3.70	3.56
25	7.77	5.57	4.68	4.18	3.85	3.63	3.46	3.32
30	7.56	5.39	4.51	4.02	3.70	3.47	3.30	3.17
35	7.42	5.27	4.40	3.91	3.59	3.37	3.20	3.07
∞	6.63	4.61	3.78	3.32	3.02	2.80	2.64	2.51

References

1. Brune P, Dyer J, Masters D. Unpublished research, University of Rochester, 2003.
2. Funkenbusch PD, Zhang X. Analysis of an adhesion test using a designed (matrix) experiment. Proceedings of the 23rd Annual Meeting of the Adhesion Society. 2000; 519–521.
3. Felba J, Friedel KP, Wojcicki S. The optimization of a triode electron gun with a thermonic cathode. Appl Surf Sci 1997; 111:126–134.
4. Amorsolo AV Jr, Funkenbusch PD, Kadin AM. A parametric study of titanium silicide formation by rapid thermal processing. J Mater Res 1996; 11:1–10.
5. Chen RS, Lee HH, Yu CY. Application of Taguchi's method on the optimal process design of an injected molded PC/PBT automobile bumper. Compos Struct 1997; 39:209–214.
6. Whitney DE, Tung ED. Robot grinding and finishing of cast iron stamping dies. Trans ASME 1992; 114:132–140.
7. Khami R, Khan A, Micek D, Zubeck J, O'Keefe M. Optimization of the conical air system's mass airflow sensor performance. Proceedings of the 12th Annual Taguchi Symposium 1994:147–160.

8. Bandurek GR, Disney J, Bendell A. Application of Taguchi Methods to Surface Mount Processes. Qual Reliab Eng Int 1998; 4:171–181.

9. El-Haiku B, Uralic J, Cecile E, Drayton M. Front end accessory drive power steering press fit functional optimization. Proceedings of the 17th Annual Taguchi Symposium 1999:81–99.

10. Reddy PBS, Nishina K, Subash BA. Performance improvement of a zinc plating process using Taguchi's methodology: a case study. Met Finish 1998:24–34.

11. Shiau Y-R, Jiang BC. Determine a vision system's 3D coordinate measurement capability using Taguchi methods. Int J Prod Res 1991; 29:1101–1122.

12. Sano H, Watanabe M, Fujiwara S, Kurihara K. Air flow noise reduction of inter-cooler system. Proceedings of the 17th Annual Taguchi Methods Symposium. Am Suppl Inst 1997:261–271.

Index

Printed in the United States
by Baker & Taylor Publisher Services